THE FUTURE OF TECHNOLOGY

OTHER ECONOMIST BOOKS

Guide to Analysing Companies
Guide to Business Modelling
Guide to Business Planning
Guide to Economic Indicators
Guide to the European Union
Guide to Financial Markets
Guide to Management Ideas
Numbers Guide
Style Guide

Dictionary of Business
Dictionary of Economics
International Dictionary of Finance

Brands and Branding
Business Consulting
Business Ethics
Business Strategy
China's Stockmarket
Globalisation
Headhunters and How to Use Them
Successful Mergers
Wall Street

Essential Director
Essential Economics
Essential Finance
Essential Internet
Essential Investment
Essential Negotiation

Pocket World in Figures

The
Economist

THE FUTURE OF TECHNOLOGY

THE ECONOMIST IN ASSOCIATION WITH
PROFILE BOOKS LTD

Published by Profile Books Ltd
3A Exmouth House, Pine Street, London EC1R 0JH

Typeset in EcoType by MacGuru
info@macguru.org.uk

Printed and bound in Great Britain by
Creative Print and Design (Wales), Ebbw Vale

A CIP catalogue record for this book is available
from the British Library

ISBN 1 86197 971 1

Contents

The authors

Tom Standage is *The Economist*'s technology editor and the author of several books on the history of science and technology, including *The Victorian Internet*, *The Turk* and *A History of the World in Six Glasses*.

Tamzin Booth is *The Economist*'s media correspondent.

Geoffrey Carr is science editor of *The Economist*.

Ben Edwards is *The Economist*'s American business editor based in New York.

Andreas Kluth is *The Economist*'s technology correspondent in San Francisco.

Natasha Loder is science correspondent at *The Economist*.

Ludwig Siegele is *The Economist*'s Berlin correspondent. He was previously technology correspondent in San Francisco.

Vijay Vaitheeswaran is *The Economist*'s energy and environment correspondent and the author of *Power to The People: How the Coming Energy Revolution Will Transform an Industry, Change Our Lives, and Maybe Even Save the Planet*.

Thanks also to Anne Schukat and Chip Walter, freelance journalists who contributed to some of the articles in this book.

Foreword

TO UNDERSTAND THE FUTURE OF TECHNOLOGY, start by looking at its past. From the industrial revolution to the railway age, through the era of electrification, the advent of mass production and finally to the information age, the same pattern keeps repeating itself. An exciting, vibrant phase of innovation and financial speculation is followed by a crash; then begins a longer, more stately period during which the technology is widely deployed. Consider the railway mania of the 19th century, the dotcom technology of its day. Despite the boom and bust, railways subsequently proved to be a genuinely important technology and are still in use today – though they are not any longer regarded as technology by most people, but as merely a normal part of daily life.

Having just emerged from its own boom and bust, the information-technology industry is at the beginning of its long deployment phase. Irving Wladawsky-Berger, a technology guru at IBM, has said that the industry has entered its "post-technology" period. His point is not that technology no longer matters, but that how it is applied is now more important than the technology itself. All successful technologies eventually fade into the background of everyday life, as electricity and railways have, becoming both more important and less visible. That is now happening to information technology. The history of technology thus provides clues about its future.

To be sure, the information-technology industry is ambivalent about its growing maturity. On the one hand, it means that the debates of the past decade – whether businesses have any choice but to embrace the internet, for example, or whether it is possible to make money online – have been resolved, and in the technologists' favour. All the hype of the internet boom contained a kernel of truth, though harnessing the new technology proved harder and is taking longer than the cheerleaders of the 1990s anticipated. That is the nature of revolutions: they are rarely as quick or as clean as the revolutionaries expect them to be.

On the other hand, however, this means that the revolutionary ideas of a few years ago have now become conventional wisdom. Having convinced the business world of the merits of technology, the industry has lost much of its iconoclastic fervour. The corporate adoption of information technology has become such old hat, indeed, that Nicholas Carr, an editor at the *Harvard Business Review*, even published an article in 2003 entitled "IT doesn't matter". Many technologists were livid, but Mr Carr had a point. While there are some areas where technology can still provide competitive advantage, its widespread adoption means that it makes less difference than it used to.

This shift has many implications for the future of technology, which will be examined in the first part of this book. For one thing, the industry needs to adjust its self-image as it moves from youthful exuberance to middle-aged pragmatism. It also needs to place more emphasis on security and reliability, as computers become central to so many aspects of business and personal life, from e-mail to online shopping and banking. Fixing security problems may be less sexy than dreaming up whizzy new technologies, but it must be done.

Another problem is that while information technology can do all sorts of clever things, it is still too complex in many respects, and places insufficient emphasis on the needs of its users. This is understandable in the early days, when the main thing is to get new technologies to work at all, but eventually it hinders their spread and limits their potential. Lastly, companies are increasingly exploiting the plunging costs of computing and communications to make their internal processes more flexible and efficient, in particular through outsourcing. This has put information technology at the heart of the battles over globalisation and development – unfamiliar ground for most technologists, who generally prefer to avoid politics.

Does this mean that technology has completely lost its shine and become a dull, tedious business? Far from it. Rather, as corporate computing has matured, the technology industry's innovative focus has moved into consumer electronics. It used to be the case that the most advanced computing and communications technologies were to be found sealed away in corporate data-centres; but new technologies now appear first in consumer gadgets, which are produced by the million. Mobile phones, nearly 2 billion of which are now in use around the world, have become the most ubiquitous form of information technology on earth, while video-game consoles have become one of the most advanced, with far more computing power than office PCs.

The growing ubiquity and sophistication of consumer-electronics devices is the topic of the second part of this book. Consumer-electronics firms are taking the fruits of the information-technology boom and applying them not to corporate systems, where they started out, but to every conceivable aspect of daily life. It is a logical consequence of the information-technology boom of the late 20th century. The first hard disks, for example, were attached to mainframes; now hard disks can be found in the living room, inside personal video recorders, or even in your pocket, inside an iPod music-player.

As well as mobile phones and games consoles, the two areas where the rapid pace of innovation is most evident, there is an industry-wide effort to interconnect consumer-electronics devices of all kinds in a seamless way, to create a "digital lifestyle" within a "digital home". It is still more of a vision than a reality, but the building blocks from which the digital home will be constructed are becoming apparent: exotic new wireless technologies, cheap storage, broadband internet access and large, flat-panel displays. All of this ought to allow music, photos, movies and games to be accessed and enjoyed with unprecedented ease – eventually. But first there are technical hurdles to overcome, new business models to devise, and legal and copyright problems to sort out. For the music and movie industries are institutionally suspicious of new technologies – despite the fact that, time after time, new inventions that they initially perceived as threats ended up expanding their businesses.

In a sense, though, the current boom in consumer technology is merely an aftershock of the information-technology boom, rather than an entirely new phase of technological development. So what comes next? Once information technology has percolated into everything, from wireless sunglasses that double as headphones to radio-tagged cereal boxes, what new technology will lead the next great phase of transformation, disruption and creativity?

The third part of this book will examine some of the leading contenders. The first is biotechnology, which promises new medical treatments tailored for individual patients, new crops and novel industrial processes. Indeed, the industrial uses of genetic modification could prove be far more significant than its better known (and highly controversial) agricultural applications. A second contender is energy technology. There have been energy revolutions in the past, notably the introduction of steam power, which spawned the industrial revolution, and the advent of electrification, which prompted further industrial transformation. As global energy consumption grows, reserves of fossil fuels decline and climate change accelerates, there is likely to be a growing need for better, cleaner energy technologies.

A third candidate is nanotechnology, the exploitation of the unusual phenomena that manifest themselves at the nanoscale (a nanometre is a thousand millionth of a metre). This emerging field has been the topic of speculation and controversy for some years, but is now moving from the realm of science fiction to the marketplace. Lastly, we consider two technologies – robotics and artificial intelligence – that have been touted as the "next big thing" in the past and are now considered failures. Nevertheless, both have arguably been quite widely adopted, which suggests that evaluating the success of new technologies is more difficult than it seems.

All three parts of this book consist of surveys and articles that appeared in *The Economist* between 2002 and 2005; the date of publication is given at the end of each one. Each article reflects the author's point of view at the time in question, but while some have been lightly edited, they have not been substantially rewritten, and are as valid today as they were at the time of publication. Collectively they illustrate how the technology industry is changing, how technology continues to affect many aspects of everyday life – and, looking further ahead, how researchers in several promising fields are developing the innovations that seem most likely to constitute the future of technology.

PART 1

INFORMATION TECHNOLOGY GROWS UP

Part 1 consists of four surveys. "Coming of age", a survey of information
technology, looks at the industry's growing maturity as it moves from
iconoclastic adolescence to pragmatic middle-age. "Securing the cloud", a
survey of digital security, examines the importance of security as digital
technology is increasingly relied upon. "Make it simple", another survey of
information technology, explains how the industry's ability to invent new
technologies has outstripped its ability to explain them to its customers,
and examines what can be done about it. Lastly, "A world of work", a
survey of outsourcing, examines the business, technical and political
issues raised by the use of technology to move work overseas.

1
COMING OF AGE

Paradise lost

So far, information technology has thrived on exponentials. Now it has to get back to earth

CLOSE YOUR EYES and think of information technology. You might picture your PC crashing yet again, and recall that your teenager was supposed to fix it. That leads you to the 12-year-old hacker who broke into a bank's computer system the other day, which brings to mind the whizz-kids in a garage inventing the next big thing that will turn them into the youngest billionaires ever.

In IT, youth seems to spring eternal. But think again: the real star of the high-tech industry is in fact a grey-haired septuagenarian. Back in 1965, Gordon Moore, co-founder of Intel, the world's biggest chipmaker, came up with probably the most famous prediction in IT: that the number of transistors which could be put on a single computer chip would double every 18 months. (What Mr Moore actually predicted was that the figure would double every year, later correcting his forecast to every two years, the average of which has come to be stated as his "law".)

This forecast, which implies a similar increase in processing power and reduction in price, has proved broadly accurate: between 1971 and 2001, transistor density has doubled every 1.96 years (see Chart 1.1). Yet this pace of development is not dictated by any law of physics. Instead, it has turned out to be the industry's natural rhythm, and has become a self-fulfilling prophecy of sorts. IT firms and their customers wanted the prediction to come true and were willing to put up the money to make it happen.

Even more importantly, Moore's law provided the IT industry with a solid foundation for its optimism. In high-tech, the mantra goes, everything grows exponentially. This sort of thinking reached its peak during the internet boom of the late 1990s. Suddenly, everything seemed to be doubling in ever-shorter time periods: eyeballs, share prices, venture capital, bandwidth, network connections. The internet mania began to look like a global religious movement. Ubiquitous cyber-gurus, framed by colourful PowerPoint presentations reminiscent of stained glass, prophesied a digital land in which growth would be limitless, commerce frictionless and democracy direct. Sceptics were derided as bozos "who just don't get it".

Today, everybody is older and wiser. Given the post-boom collapse of spending on IT, the idea of a parallel digital universe where the laws of economic gravity do not apply has been quietly abandoned. What has yet to sink in is that this downturn is something more than the bottom of another cycle in the technology industry. Rather, the sector is going through deep structural changes which suggest that it is growing up or

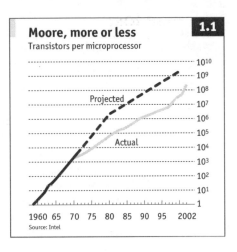

Moore, more or less
Transistors per microprocessor

10^{10}
10^9
10^8
10^7
Projected
10^6
10^5
Actual
10^4
10^3
10^2
10^1
1

1960 65 70 75 80 85 90 95 2002
Source: Intel

1.1

even, horrors, maturing. Silicon Valley, in particular, has not yet come to grips with the realities, argues Larry Ellison, the chief executive of Oracle, a database giant (who in his early 60s still sports a youthful hairdo). "There's a bizarre belief that we'll be young forever," he says.

It is not that Moore's law has suddenly ceased to apply. In fact, Mr Moore makes a good case that Intel can continue to double transistor density every 18 months for another decade. The real issue is whether this still matters. "The industry has entered its post-technological period, in which it is no longer technology itself that is central, but the value it provides to business and consumers," says Irving Wladawsky-Berger, a senior manager at IBM and another grey-haired industry elder.

Scholars of economic history are not surprised. Whether steam or railways, electricity or steel, mass production or cars – all technological revolutions have gone through similar long-term cycles and have eventually come of age, argues Carlota Perez, a researcher at Britain's University of Sussex, in her book *Technological Revolutions and Financial Capital: The Dynamics of Bubbles and Golden Ages* (Edward Elgar, 2002).

In her model (see Chart 1.2 overleaf), technological revolutions have two consecutive lives. The first, which she calls the "installation period", is one of exploration and exuberance. Engineers, entrepreneurs and investors all try to find the best opportunities created by a technological big bang, such as Ford's Model T in 1908 and Intel's first microprocessor in 1971. Spectacular financial successes attract more and more capital, which leads to a bubble. This is the "gilded age" of any given technology, "a great surge of development", as Ms Perez calls technological revolutions.

5

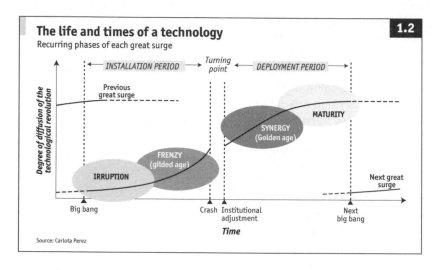

The life and times of a technology `1.2`
Recurring phases of each great surge

Source: Carlota Perez

The second, or "deployment", period is a much more boring affair. All the quick bucks have been made, so investors prefer to put their money into the real economy. The leading firms of the new economy become bigger and slower. The emphasis is no longer on raw technology, but on how to make it easy to use, reliable and secure. Yet this period is also the "golden age" of a technology, which now penetrates all parts of society.

These two periods of a technological revolution are separated by what Ms Perez calls a "turning point" – a crucial time for making the choices that determine whether a technological revolution will deliver on its promises. In her book, she concentrates mainly on the social and regulatory decisions needed to allow widespread deployment of new technology. But the same argument applies to technology vendors and customers. To enter their "golden age", they have to leave their youthful excesses behind and grow up.

A duller shade of gold

This section will examine how much grey the IT industry (and its leaders' hair) has already acquired. The first three articles are about technological shifts, and how value is moving from the technology itself to how it is applied. Many of the wares that made the IT industry's fortunes in the installation period are becoming a commodity. To overcome this problem, hardware vendors are developing new software that allows networks of machines to act as one, in effect turning computing into a utility. But the IT industry's most profitable layer will be services of all kinds, such as software delivered as an online service, or even business consulting.

The second half of this section looks at institutional learning, which has caused the value created by the IT industry to be increasingly captured by its customers. For the first time in its history, the IT industry is widely adopting open standards. Equally important, buyers are starting to spend their IT budgets more wisely. Meanwhile, the industry's relationship with government is becoming closer.

All this suggests that the technology industry has already gone greyish at the temples since the bubble popped, and is likely to turn greyer still. Sooner or later the sector will enter its "golden age", just as the railways did. When Britain's railway mania collapsed in 1847, railroad shares plunged by 85%, and hundreds of businesses went belly-up. But train traffic in Britain levelled off only briefly, and in the following two decades grew by 400%.

So are the IT industry's best days yet to come? There are still plenty of opportunities, but if the example of the railways is anything to go by, most IT firms will have to make do with a smaller piece of the pie. As this newspaper (then called The Economist, Weekly Commercial Times, Bankers' Gazette, and Railway Monitor) observed in 1857: "It is a very sad thing unquestionably that railways, which mechanically have succeeded beyond anticipation and are quite wonderful for their general utility and convenience, should have failed commercially."

Brad DeLong, an economics professor at the University of California at Berkeley, puts it somewhat more succinctly: "I am optimistic about technology, but not about profits."

Modifying Moore's law

Many of the innovations that made the IT industry's fortunes are rapidly becoming commodities – including the mighty transistor

IF GOOGLE WERE TO CLOSE down its popular web-search service tomorrow, it would be much missed. Chinese citizens would have a harder time getting around the Great Firewall. Potential lovers could no longer do a quick background check on their next date. And college professors would need a new tool to find out whether a student had quietly lifted a paper from the internet.

Yet many IT firms would not be too unhappy if Google were to disappear. They certainly dislike the company's message to the world: you do not need the latest and greatest in technology to offer outstanding services. In the words of Marc Andreessen of Netscape fame, now chief executive of Opsware, a software start-up: "Except applications and services, everything and anything in computing will soon become a commodity."

Exactly what is meant by "commoditisation", though, depends on whom you talk to. It is most commonly applied to the PC industry. Although desktops and laptops are not a truly interchangeable commodity such as crude oil, the logo on a machine has not really mattered for years now. The sector's most successful company, Dell, is not known for its technological innovations, but for the efficiency of its supply chain.

As the term implies, "commoditisation" is not a state, but a dynamic. New hardware or software usually begins life at the top of the IT heap, or "stack" in geek speak, where it can generate good profits. As the technology becomes more widespread, better understood and standardised, its value falls. Eventually it joins the sector's "sediment", the realm of bottom feeders with hyper-efficient metabolisms that compete mainly on cost.

Built-in obsolescence

Such sedimentation is not unique to information technology. Air conditioning and automatic transmission, once selling points for a luxury car, are now commodity features. But in IT the downward movement is much faster than elsewhere, and is accelerating – mainly thanks to

Moore's law and currently to the lack of a new killer application. "The industry is simply too efficient," says Eric Schmidt, Google's chief executive (who seems to have gone quite grey during his mixed performance at his previous job as boss of Novell, a software firm).

The IT industry also differs from other technology sectors in that its wares become less valuable as they get better, and go from "undershoot" to "overshoot," to use the terms coined by Clayton Christensen, a professor at Harvard Business School. A technology is in "undershoot" when it is not good enough for most customers, so they are willing to pay a lot for something that is a bit better although not perfect. Conversely, "overshoot" means that a technology is more than sufficient for most uses, and margins sink lower.

PCs quickly became a commodity, mainly because IBM outsourced the components for its first venture into this market in the early 1980s, allowing others to clone the machines. Servers have proved more resistant, partly because these powerful data-serving computers are complicated beasts, partly because the internet boom created additional demand for high-end computers running the Unix operating system.

But although expensive Unix systems, the strength of Sun Microsystems, are – and will probably remain for some time – a must for "mission-critical" applications, servers are quickly commoditising. With IT budgets now tight, firms are increasingly buying computers based on PC technology. "Why pay $300,000 for a Unix server," asks Mr Andreessen, "if you can get ten Dell machines for $3,000 each – and better performance?"

Google goes even further. A visit to one of the company's data centres in Silicon Valley is a trip back to the future. In the same way that members of the Valley's legendary Homebrew Computer Club put together the first PCs using off-the-shelf parts in the early 1970s, Google has built a huge computer system out of electronic commodity parts.

Modern Heath Robinsons

When the two Stanford drop-outs who founded Google, Sergey Brin and Larry Page, launched the company in 1998, they went to Fry's, an electronics outlet where the Valley's hardcore computer hobbyists have always bought their gear. Even today, some of the data centres' servers appear to be the work of tinkerers: circuit boards are sagging under the weight of processors and hard disks, and components are attached by Velcro straps. One reason for the unusual design is that parts can be easily swapped when they break. But it also allows Google's servers to

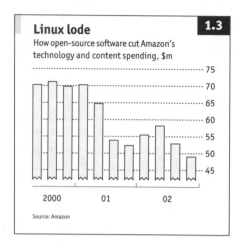

Linux lode **1.3**

How open-source software cut Amazon's
technology and content spending, $m

Source: Amazon

be made more powerful without having to be replaced completely.

What makes it easier for Google to swap off-the-shelf components is that much of its software is also a commodity of sorts. Its servers run Linux, the increasingly popular open-source operating system developed by a global community of volunteer programmers, and Apache, another open-source program, which dishes up web pages.

Because Google has always used commodity hardware and software, it is not easy to calculate how much money it has saved. But other firms that have recently switched from proprietary gear say they have significantly reduced their IT bill. Amazon.com, the leading online shopping mall, for instance, managed to cut its quarterly technology spending by almost $20m (see Chart 1.3).

The most interesting feature of Google's data centre, however, is that its servers are not powered by high-end chips, and probably will not have Itanium, Intel's most powerful processor, inside for some time yet, if ever. This sets Google apart among hot Silicon Valley start-ups, whose business plans are mostly based on taking full advantage of the exponential increase in computing power and similar growth in demand for technology.

"Forget Moore's law," blared the headline of an article about Google in *Red Herring*, a now-defunct technology magazine. That is surely overblown, but Google's decision to give Itanium a miss for now suggests that microprocessors themselves are increasingly in "overshoot", even for servers – and that the industry's 30-year race for ever more powerful chips with smaller and smaller transistors is coming to an end.

Instead, other "laws" of the semiconductor sector are becoming more important, and likely to change its underlying economics. One is the fact that the cost of shrinking transistors also follows an exponential upward curve. This was no problem as long as the IT industry gobbled up new chips, thus helping to spread the cost, says Nick Tredennick, editor of the *Gilder Technology Report*, a newsletter. But now, argues Mr Tredennick, much of the demand can be satisfied with "value transistors" that offer adequate performance for an application at the lowest possible cost, in the same way as Google's. "The industry has been focused on Moore's law because the transistor wasn't good enough," he says. "In the future, what engineers do with transistors will be more important than how small they are."

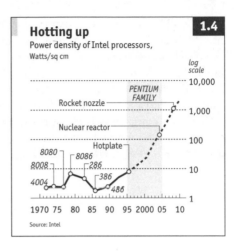

Hotting up   **1.4**
Power density of Intel processors, Watts/sq cm

log scale

PENTIUM FAMILY
Rocket nozzle
Nuclear reactor
Hotplate
8080
8086
8008
286
4004
386
486

10,000
1,000
100
10
1

1970 75   80   85   90   95 2000 05   10

Source: Intel

This is nothing new, counters Paul Otellini, Intel's chief executive. As chips become good enough for certain applications, new applications pop up that demand more and more computing power, he says: once Google starts offering video searches, for instance, it will have to go for bigger machines. But in recent years, Intel itself has shifted its emphasis somewhat from making ever more powerful chips to adding new features, in effect turning its processors into platforms.

In 2003 it launched Centrino, a group of chips that includes wireless technology. The Centrino chips are also trying to deal with another, lesser-known, limiting factor in chipmaking: the smaller the processors become, the more power-hungry and the hotter they get (see Chart 1.4). This is because of a phenomenon called leakage, in which current escapes from the circuitry. The resulting heat may be a mere inconvenience for users of high-end laptops, who risk burning their hands or thighs, but it is a serious drawback for untethered devices, where it shortens battery life – and increasingly for data centres as well, as Google again shows.

Cool chips

The firm's servers are densely packed to save space and to allow them to communicate rapidly. The latest design is an eight-foot rack stuffed with 80 machines, four on each level. To keep this computing power-house from overheating, it is topped by a ventilation unit which sucks air through a shaft in its centre. In a way, Google is doing to servers

11

what Intel has done to transistors: packing them ever more densely. It is not the machines' innards that count, but how they are put together.

Google has thus created a new computing platform, a feat that others are now replicating in a more generalised form. Geoffrey Moore (no relation), chairman of the Chasm Group, a consultancy, and a partner at Mohr, Davidow Ventures, a Silicon Valley venture-capital firm, explains it this way: computing is like a game of Tetris, the computer-game classic; once all the pieces have fallen into place and all the hard problems are solved, a new playing field emerges for others to build on.

Moving up the stack

The network is becoming the computer – and the IT industry's dominant platform

COMPUTING IS SUPPOSED TO BE the ultimate form of automation, but today's data centres can be surprisingly busy with people. When an application has to be updated or a website gets more visitors than expected, system administrators often have to install new programs or set up new servers by hand. This can take weeks and often turns out to be more complicated than expected.

Google's data centres, however, look deserted most of the time, with only about 30 employees to look after a total of 54,000 servers, according to some estimates. This is in part because machines doing searches need less care than those running complex corporate applications; but more importantly, the firm's programmers have written code that automates much of what system administrators do. It can quickly change a computer that sifts through web pages into a server that dishes up search results. Without the program, Google would have to hire many more people.

It all goes to show that another law in computing, proclaimed by Gordon Bell, another greying industry legend, still holds true: in IT, the dominant platform shifts every ten years or so. Mainframes, minicomputers, PCs and servers are now likely to be followed by a grid of computers, either within a data centre or as a disparate collection of connected machines. The network will at last be the computer, to paraphrase a slogan coined by Sun Microsystems. Machines will no longer simply be attached to a network: instead, the network will allow them to act as one.

Yet this new platform, which computer scientists like to call "grid computing", is less about replacing old technology and more about managing the existing gear – another sign that IT is maturing. Merrill Lynch's Steve Milunovich, one of the leading hardware analysts on Wall Street, says that IT has entered the era of "managed computing". Forrester Research, a high-tech consultancy, has coined the term "organic IT" – a computing infrastructure that is not only built on cheap parts, but is also as adaptive as a living organism. Whatever label the industry settles for, the race to lead in the next round of computing is already on. The new platform gives those threatened by commoditisation a chance to differentiate themselves by moving up the technology stack to a potentially more lucrative layer.

There is every incentive for HP, IBM, Microsoft and Sun, as well as a raft of start-ups, to encourage this shift, but there is also a real need for a new platform. Computing has certainly got faster, smarter and cheaper, but it has also become much more complex. Ever since the orderly days of the mainframe, which allowed tight control of IT, computer systems have become ever more distributed, more heterogeneous and harder to manage.

Managing complexity

In the late 1980s, PCs and other new technologies such as local area networks (LANs) allowed business units to build their own systems, so centralised IT departments lost control. In the late 1990s, the internet and the emergence of e-commerce "broke IT's back", according to Forrester. Integrating incompatible systems, in particular, has become a big headache.

A measure of this increasing complexity is the rapid growth in the IT services industry. According to some estimates, within a decade 200m IT workers will be needed to support a billion people and businesses connected via the internet. Managing a storage system already costs five times as much as buying the system itself, whereas less than 20 years ago the cost of managing the system amounted to only one-third of the total (see Chart 1.5).

What is more, many of today's IT systems are a patchwork that is inherently inefficient, so firms spend 70–90% of their IT budgets simply on keeping their systems running. And because those systems cannot adapt quickly to changes in demand, companies overprovide. They now spend almost $50 billion a year on servers, but the utilisation rate for these computers is often below 30%.

Besides, complexity is bound to increase, predicts Greg Papadopoulos, Sun's chief technology officer. Today, the electronics to hook up any device to the network cost about $1. In ten years' time, the price will be down to one cent. As a result, he says, the number of connected things will explode, and so will the possible applications. For example, it will become practical to track items such as razor blades (10% of which apparently disappear on their way from the factory to the shop).

When things get too complicated, engineers usually add a layer of code to conceal the chaos. In some ways, the current shift in computing is the equivalent of what happened when cars became easy to use and drivers only had to turn the key instead of having to hand-crank the engines. In geek speak, adding such a new layer is called "raising the level of abstraction". This happened when PC operating systems first hid the nuts and bolts of these computers and gave them a simple user inter-

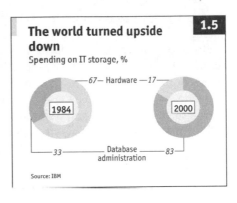

The world turned upside down

Spending on IT storage, %

— 67— Hardware —17—

1984

2000

—33— Database administration —83—

Source: IBM

1.5

face, and it is happening now with the new platform, which is already being compared to an operating system for data centres or computing grids.

Just like Google's management program, this grid computing software (only half-jokingly called "griddleware" by some) automates much of the work of system administrators. But it is also supposed to serve a higher purpose: "virtualisation". Put simply, this means creating pools of processing power, storage capacity and network bandwidth. A data centre, or a collection of machines on a network, thus becomes a virtual computer whose resources can be allocated as needed. The ultimate goal is that managed computing will become rather like flying a modern jet plane: IT workers will tell the system what kind of applications it should run, and then deal only with exceptions.

Although the rivals in this new field are pretty much on the same technological track, their strategies are different. Some of the numerous start-ups already have working products – and no hidden agenda, says Mr Andreessen, of Opsware, the leading newcomer: "We don't need to push our customers also to buy other stuff from us." The incumbents, on the other hand, want the new software layer to protect their old business models as well. HP's Utility Data Centre (UDC) initiative and Sun's N1 plan are supposed to help these firms sell their profitable hardware. IBM's "autonomic computing" effort goes hand-in-hand with Big Blue's IT services business. And Microsoft's Dynamic Services Initiative (DSI) is tightly linked with its Windows operating system.

Yet despite such arm-twisting, customers are unlikely to bet solely on newcomers. Only the biggest vendors will really be able to deliver managed computing, argues Shane Robinson, the chief technology officer of HP, which has much riding on the new platform. According to the

Gartner Group, a consultancy, HP is leading in virtualisation, and views management software as its big opportunity.

Once thing is clear: once all the technical challenges of grid computing have been overcome, hardware will have become a true commodity. Machines, storage devices and networks will lose their identity and feed into pools of resources that can be tapped as needed. This liquefaction of hardware, in turn, will allow computing to become a utility, and software a service delivered online.

Techniques, not technology

IT firms hope to turn the dismal science into a profitable business

TO MOST PEOPLE the world is round, but geeks often see it as a stack of layers. In corporate computing, it starts with the hardware, on top of which sits the operating system, then the database, the applications and finally IT services. When their layer is getting commoditised, technology companies tend to move up the stack, where more money can be made.

In their quest for greener pastures, IT firms have reached new heights by moving into cutting-edge economics. Both HP and IBM have opened labs to conduct research on the subject, in the hope that this will help them to offer their customers more sophisticated services.

To be sure, economics has had its place in the IT industry for some years now. HP, for instance, already uses software that simulates markets to optimise the air-conditioning systems in its utility data centres. And IBM's Institute for Advanced Commerce has studied the behaviour of bidding agents, in the hope of designing them in such a way that they do not engage in endless price wars.

Now HP is reaching even higher, with experimental economics. As the name implies, researchers in this field set up controlled experiments with real people and real money to see whether economic theories actually work. Perhaps surprisingly, it seems that they do, as demonstrated by the work of Vernon Smith of George Mason University in Virginia. (Mr Smith is considered the founding father of this field and won the 2002 Nobel prize in economics.)

Secret agent

HP goes further. The firm's team of five researchers does not test economic theories, but tries to create "novel mechanisms to improve the fluidity of interactions in the information economy", says Bernardo Huberman, head of the group. In everyday language, the researchers are working on clever tools that make it easier to negotiate online, establish reputations and make forecasts.

Mr Huberman's group already has something to show for its efforts. It has developed a methodology for predicting uncertain events using a small group of individuals. First, they find out about their subjects' attitudes towards risk and their ability to forecast a given outcome. They then use this information to weight and aggregate their predictions of events, resulting in fairly accurate forecasts.

These tools will first be used inside the company. The top management of one of HP's divisions is already testing the forecasting methodology to predict its revenue. But ultimately the firm wants to find outside customers for its research findings. American intelligence agencies, such as the CIA, have already shown interest. They need better tools to weigh the opinions of those who analyse incoming information.

So at what point will firms such as HP and IBM have moved far enough up the stack to cease to be traditional IT vendors and become service providers or consultancies? Most analysts agree that this metamorphosis is still some way off. But it already seems certain that in future IT firms will increasingly be in the business of techniques rather than technology.

At your service

Despite early failures, computing will eventually become a utility

MARC BENIOFF (not a grey hair in sight) is not afraid to mix religion and business. In February 2003, the practising Buddhist and chief executive of salesforce.com, a San Francisco start-up, invited 200 customers and friends to a benefit concert featuring David Bowie, with proceeds going to the Tibet House, a New York cultural centre whose patron is the Dalai Lama. But Mr Benioff also used the event to get his firm's message across: "Freedom from software".

The unusual mixture raised some eyebrows, but in a way Mr Benioff's technology does indeed dispense with software. Clients can access the firm's service via a web browser, which saves them having to install a complicated customer relationship management (CRM) program on their own computers. Some 6,300 customers in 110 countries have already signed up for it, generating $52m in the 2002/03 fiscal year, says Mr Benioff.

Sceptics maintain that salesforce.com is not so much the leader of a new trend as a lone survivor of better times. Hundreds of application service providers (ASPs) were launched late in the dotcom boom, but few others have succeeded. This is mainly because with existing technology it is difficult to make money on a high-end software service.

But not for much longer. Thanks to the technological trends outlined on pages 8–16, computing is becoming a utility and software a service. This will profoundly change the economics of the IT industry. "The internet spells the death of the traditional software business model," predicts Mr Benioff.

This is not as earth-shattering as it sounds. As other technologies matured, buyers were given more choice in how to acquire them, says IBM's Irving Wladawsky-Berger. In the early days of electricity, for instance, most firms had to have their own generators. Now most can get their power from the grid. Similarly, he says, it would be surprising if in 20 years' time most of IT was not outsourced.

Traditionally, companies wanting to invest in computer systems did not have much choice: they had to build and operate their own. To be sure, they could outsource the work to companies such as EDS and IBM Global Services, but in technical terms that did not change much, because such firms usually operate dedicated computer systems for each customer.

There must be a better way

When it comes to enterprise software, in particular, this way of delivering technology creates a somewhat perverse set of economics. Software is a service at heart, albeit an automated one, but it is sold much like a manufactured good. Customers have to pay large sums of money up front, bear much of the risk that a program may not work as promised and cannot readily switch vendors.

IT firms, for their part, have to spend a lot of resources on marketing and distribution, rather than concentrating on developing software that works well and is easy to use. Network effects and Wall Street make matters worse. In many markets it is a great advantage to be first, so vendors are tempted to release programs even if they are still riddled with bugs. And because equity analysts rightly consider software firms a risky investment, such firms must grow quickly to justify their relatively high share prices, pushing them to sell more programs than customers need.

All this explains several peculiarities of the software business. One is the fact that many of the licences sold are never used, a phenomenon known as "shelfware". More importantly, many software firms have grown so fast, often mortgaging the future, that they collapse when they reach $1 billion in annual revenues, sometimes never to recover. Then there is the end-of-quarter rush, spurring many firms to do anything to get deals signed and meet analysts' expectations.

The need to grow quickly also explains why IT industry leaders are such a "truly extraordinary cast," in the words of Louis Gerstner, who was IBM's chief executive for eight years. "They make outrageous remarks, they attack one another publicly with great relish," he writes in his book *Who Says Elephants Can't Dance?* (HarperBusiness 2002). Bosses of software firms, in particular, need to demonstrate that they will pursue growth at almost any price – which explains why they are often paired with grey-haired chief financial officers as a calming counterweight.

Mr Gerstner, who has spent most of his career outside the IT industry, does not point fingers, but examples of the industry's "bizarre practices", as he puts it, are not hard to find. The most obvious one is Oracle, a database giant which had a near-death experience in 1991, having cut some reckless deals to meet expectations. The firm is also known to have released software prematurely, most recently its e-business suite. And it is run by Larry Ellison, arguably the most colourful boss in the industry, and Jeff Henley, the archetype of a grandfatherly CFO. To be fair, it must be said that the company has matured greatly in recent years.

In future, technology itself could lead to a better balance in the sector as a whole. The internet made it possible to run ASPs such as salesforce.com, but it also enabled hardware-makers to monitor servers and bill customers remotely on the basis of the average use per month. This is the sort of thing HP does with its high-end Superdome machines.

As data centres become automated, computing will increasingly turn into a true utility. With the management software described on pages 13–16, firms can share computing resources, which means they always have enough of them but pay only for what they actually use. They no longer need to run their own dedicated machines, any more than they need to run their own power plants.

Waiting for web services

For software truly to become a service, however, something else has to happen: there has to be wide deployment of web services. These are not, as the term might suggest, web-based offerings such as salesforce.com, but a standard way for software applications to work together over the internet. Google, for instance, also offers its search engine as a web service to be used in other web offerings, such as Googlefight, a site where surfers with time to waste can find out which of two related key words produces more search results.

Ultimately, experts predict, applications will no longer be a big chunk of software that runs on a computer but a combination of web services; and the platform for which developers write their programs will no longer be the operating system, but application servers. These are essentially pieces of software that offer all the ingredients necessary to cook up and deliver a web service or a web-based service such as salesforce.com.

Just as with management software for data centres, vendors are already engaged in a battle for dominance. Ranged on one side is Microsoft with its .NET platform (although it has recently toned down the use of this name). Jostling on the other are BEA, IBM, Oracle and Sun, with different versions of technology based on the Java programming language.

Both camps are likely to coexist, but the economics of software services will be different. Most important, vendors will be much more motivated than previously to keep their promises. "In the old world, we didn't care if you were up and running, we only cared about the numbers," says Mr Benioff, who cut his professional teeth at Oracle. "Now, I get paid only if my customers are happy."

A different kind of grey

Shifting more of the implementation risk to vendors will profoundly change the nature of the software business. Wall Street will have to view software firms more like utilities, which tend to grow rather slowly if steadily. And, perish the thought, software bosses could get more boring. The tone in the industry may no longer be set by people such as Mr Ellison, but by more prudent and cerebral chief executives such as SAP's Henning Kagermann.

Incumbents will not find it easy to manage this transition: they will have to wean themselves from the heroin of growth. Of the industry heavyweights, Oracle has arguably travelled farthest, having put most of its programs online as early as 1998. As yet, this part of its business contributes only a tiny share of the total revenue, but Mr Ellison expects it to grow quickly. He also acknowledges that the time for visionary leaders like himself may well be over: "It is mainly going to be about execution."

On the vision front, however, IBM has recently bested Mr Ellison. In October 2002, Samuel Palmisano, the firm's chief executive, announced that IBM was making a $10-billion bet on what he called "on-demand computing" – essentially an effort to turn IT from a fixed into a variable cost. American Express has already signed a seven-year, $4 billion contract with IBM which allows the credit-card company to pay only for the IT resources it needs.

Yet the American Express contract still looks more like a classic outsourcing deal with flexible pricing. If computing is to become truly on-demand, much remains to be done, says Mr Wladawsky-Berger, who leads IBM's initiative. Getting the technology right is probably the easy part. The more difficult problem is persuading the industry to settle on open standards.

The fortune of the commons

For the first time, the IT industry is widely adopting open standards – thanks to the internet

B UYING A SCREW IS EASY TODAY, if you know what kind you want. But in America in the middle of the 19th century, such a purchase could get quite complicated. Most screws, nuts and bolts were custom-made, and products from different shops were often incompatible. The craftsmen who made them liked it this way, because many of their customers were, in effect, locked in.

Yet it was one of these craftsmen's leaders who set America's machine-tool industry on the path of standardisation. In 1864, William Sellers proposed a "uniform system of screw threads", which later became widely adopted. Without standardised, easy-to-make screws, Mr Sellers' argument went, there could be no interchangeable parts and thus no mass production.

Not every technology sector had such far-sighted leaders. But railways, electricity, cars and telecommunications all learned to love standards as they came of age. At a certain point in their history, it became clear that rather than just fighting to get the largest piece of the pie, the companies within a sector needed to work together to make the pie bigger.

Without standards, a technology cannot become ubiquitous, particularly when it is part of a larger network. Track gauges, voltage levels, pedal functions, signalling systems – for all of these, technical conventions had to be agreed on before railways, electricity, cars and telephones were ready for mass consumption. Standards also allow a technology to become automated, thus making it much more reliable and easier to use.

Today, the IT industry is finally getting the standards religion. In fact, standards have always played an important role in high-tech, but they were often proprietary. "For the first time, there are true standards to allow interoperability – de jure standards not controlled by a vendor," points out Steve Milunovich, an analyst at Merrill Lynch.

This is not simply a question of protocols and interfaces. Entire pieces of software are becoming open standards of sorts. Operating systems, for instance, are technically so well understood that they can be

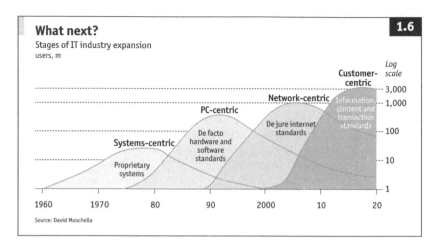

What next?

1.6

Stages of IT industry expansion

users, m

Log scale

Customer-centric

Network-centric

PC-centric

Information, content and transaction standards

De jure internet standards

De facto hardware and software standards

Systems-centric

Proprietary systems

3,000

1,000

100

10

1

1960   1970   80   90   2000   10   20

Source: David Moschella

developed by worldwide virtual communities of volunteer program-mers, as with Linux, the most popular piece of open-source software.

The taming of the screw

It would be hard to overestimate the importance of this shift. So far, just as in the early days of the screw, the name of the game in IT has been locking in customers, making it costly for them to switch from one brand of technology to another. In some ways, although IT firms are the epitome of mass production, when it comes to standards they are still stuck in the craftsmen era, which explains in large part why they have been so amazingly profitable.

Network effects make it even more attractive to control a technology, argue Carl Shapiro and Hal Varian, two economics professors, in *Information Rules*, still the best read on the network economy (Harvard Business School Press, 1998). If the value of a technology depends not just on its quality but also on the number of users, positive feedback can help one firm to dominate the market. For example, the more people are already connected to a data network using a particular transmission standard, the more people will see the point of hooking up to it.

These network effects also explain why the IT industry in the 1980s already started to move away from completely proprietary technology, the hallmark of the mainframe era. Microsoft, in particular, figured out how to strengthen feedback loops by encouraging other software firms to develop applications for its operating system. This kind of openness made Windows a standard, but users were still locked in.

Now it seems that, thanks to the internet, the IT industry has entered a positive feedback loop in favour of open standards. Looking back, says Mr Wladawsky-Berger, historians will say that the internet's main contribution was to produce workable open standards, such as TCP/IP, its communication protocol, or HTML, the language in which web pages are written. The internet has also made it much easier to develop standards. Most of the work in the Internet Engineering Task Force (IETF) and the World Wide Web Consortium (W3C), the internet's main standards bodies, is done online. Global open-source communities are able to function because their members can communicate at almost no cost using e-mail or other online tools.

The success of these groups has also inspired traditional IT companies to create their own open-source-like bodies. Sun, for instance, launched the "Java Community Process", or JCP, to develop its Java technology. But because Sun is worried that its standard could splinter, just as that for the Unix operating system did, the firm has installed itself as the JCP's benevolent dictator.

Sun is not the only firm to have learned that creating standards can be good for business – for instance, to commoditise a complementary good or to prevent a single firm from controlling an important technology. If operating systems become more of a commodity, reason IBM and others who back Linux, this will make customers spend more money on other products and weaken both Microsoft and Sun.

A new incentive

The emergence of web services has concentrated minds wonderfully on developing open standards. Displaying an unprecedented degree of co-operation, the computer industry is developing a host of common technical rules that define these new kinds of online offerings. Hence the proliferation of new computer-related acronyms such as XML, SOAP, UDDI, WSDL and so on.

To be sure, standardising web services is not always easy. As standardisation moves into more complex areas, such as security and the co-ordination of different offerings, consensus seems to be harder to achieve. Incumbents in particular have started to play games to give their wares an advantage. They are also trying to lock in customers by adding proprietary extensions to the standards mix.

Most worrying, however, is the possibility that software firms will have to pay if they implement web-services standards. Most standards bodies currently allow firms to continue owning the intellectual property they contribute as long as they do not charge for it. But the more involved that standards for web services become, the greater the pressure that firms should be able to charge for the use of the patents they have invested in.

Smaller web-services firms have already started ringing the alarm bells. The IT industry is at a crossroads, says Eric Newcomer, chief technology officer of Iona Technologies. One road leads to a truly standardised world in which companies are able to reap all the benefits of web services. The other road "leads back to yesteryear, where proprietary systems ruled the day".

The controversy points to a more general problem with technology standards: where to draw the line between the IT commons and the areas where firms should compete with proprietary technology. If the commons area is too large, there might not be enough incentive to innovate. If it is too small, incompatibilities could keep web services from becoming a standard way for computer systems to communicate.

This dividing line is flexible, particularly when it comes to something as malleable as software. But in the long run, says Ken Krechmer, a telecommunications-standards expert, information technology itself will help to reconcile standardisation and innovation, because it will increasingly turn standards into "etiquettes".

Systems such as railways or electricity and telephone networks, Mr Krechmer argues, need "compatibility" standards – clear specifications on how they can interoperate. But information technology is "adaptive", meaning that as devices become more intelligent, they can negotiate which standard they want to use to communicate. What is needed is a "meta-protocol", regulating the back and forth.

Faxes already work this way. Before transmitting anything, they negotiate over the speed at which they want to communicate. The extensible markup language (XML), the lingua franca underlying most web-services standards, also enables etiquettes. If the computer systems of two companies want to exchange the XML document for an order, they can first come to a common understanding of what the file's information means. Etiquettes thus allow for proprietary innovation while ensuring compatibility, argues Mr Krechmer.

The customer is king

In the end, though, how proprietary or how open the IT industry is likely to be will depend on its customers – who seem increasingly keen on open standards. "Vendors no longer lock in customers," says Robert

Gingell at Sun. "Now it's customers locking in vendors – by telling them which standards they have to support."

What is more, customers themselves are making their voices heard more clearly in the standards-setting process. The Liberty Alliance, an industry group developing specifications on how to manage identities and personal information online, was originally launched by Sun as a counterweight to Microsoft's Passport service, but is now driven by large IT users such as United Airlines, American Express and General Motors.

And it is not just because they hate to get locked in that customers get involved. Increasingly, says William Guttman, an economics professor at Carnegie Mellon University, standards must take account of public-policy issues such as privacy. Without the input of users, governments and academics, as well as IT firms and their customers, specifications risk becoming irrelevant, Mr Guttman maintains. He himself has launched an inclusive group called the Sustainable Computing Consortium (SCC), which among other things is looking for ways of measuring software quality.

Customers, in short, are getting more sophisticated all round – but most notably when it comes to investing in IT.

Cold killer application

The IT industry's customers are demanding more bang for fewer bucks

THE INTERNET BUBBLE and the subsequent accounting scandals had at least one merit: most people now know what chief information and chief financial officers do. In years to come, they will have to get used to a combination of both jobs: the CFO of IT.

Yet for now, hardly anybody has heard of such a thing. Marvin Balliet, whose official title at Merrill Lynch is "CFO, Global Technology & Services", says that even internally he has a lot of explaining to do. Simply put, his job is to make sure that the bank's annual IT budget of more than $2 billion is wisely spent. This means bridging two worlds: IT people on the one hand and business units on the other. The business people need to know what is technically possible, and the IT lot what is financially feasible.

Mr Balliet, and the growing number of managers with similar titles, are living proof that technology buyers too are on a steep learning curve. Companies that invested recklessly during the bubble years, and stopped when it burst, are at last getting ready to make more rational technology decisions. "Capitalism has made its entry into IT," says Chris Gardner, co-founder of iValue, a consultancy, and author of *The Valuation of Information Technology* (John Wiley, 2000).

Yet this is not just a predictable reaction to the boom-and-bust cycle. There is big money at stake. After almost 40 years of corporate IT, technology investment now often makes up more than half of capital spending. As John O'Neil, chief executive of Business Engine, a project-management firm, puts it: "IT can't hide any more."

Why should it have wanted to hide in the first place? Part of the reason is that IT projects are usually highly complex affairs that change constantly and tend to get out of control. "Traditionally, managing technology was magic, with quality and performance delivered only through incredible feats of highly skilled people," says Bobby Cameron, who cut his professional teeth in the days of punch cards and is now an analyst with Forrester Research.

Even today, IT departments, particularly in America, are often magic kingdoms full of technology wizards where basic business rules do not seem to apply. Investment decisions are generally guided by gut feeling and by the latest wheeze, rather than by the firm's overall business strategy and sound financial analysis.

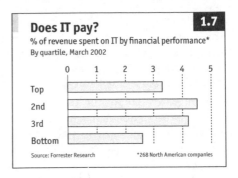

Does IT pay? `1.7`
% of revenue spent on IT by financial performance*
By quartile, March 2002

Source: Forrester Research   *268 North American companies

This is not just the fault of IT people who cherish their role as lone gurus, but also of their bosses who often abdi-cate responsibility to technologists and set no clear rules on how to make decisions. Business units, for their part, often start too many pro-jects and do not take responsibility for their success or failure. After all, in most companies, IT costs are not directly allocated to those who incur them.

Whose job?

This set-up creates permanent tension between the IT departments and the business units, which is why most CIOs do not survive in their jobs for more than two years. It is also the main reason why so many IT pro-jects are over budget and late. And when they are up and running at last, they often turn out to be obsolete already; or they do not get used because they take no account of how employees actually do their work.

High-tech consultancies estimate that more than half of all IT projects go wrong. They may have an incentive for exaggerating the failure rate (the more problems there are, the greater the perceived need for consul-tancy), but there is no question that much IT investment is wasted. To complicate matters, firms appear to differ widely in how efficiently they invest in IT. Looking at the relationship between the technology budgets and the financial results of 268 American firms, Forrester found that those that spend the most on IT are not necessarily the best performers (see Chart 1.7).

Such statistics, along with their own unhappy experiences, have led many firms to rethink the way they spend their IT dollars. Using com-plex valuation methods, they try to work out beforehand whether IT projects are likely to return the investment. "They now have to com-pete for capital with other forms of spending," says Chris Lofgren, chief executive of Schneider National, an American trucking and logis-tics company.

The trend has already generated a cottage industry for tools to calculate return on investment (ROI) and similar financial measures. One of the most sophisticated of these is offered by iValue. This start-up assesses all possible effects of an IT project – from customer loyalty and likely adoption rates to the company's share price – by building complex economic simulations for its customers, which include Citibank's Global Securities arm and Baan, a software firm.

Other tools allow firms to budget for IT projects, keep track of them and allocate the costs. Business Engine operates a web-based service that brings together all the information about a project and allows everybody involved to collaborate. One of the reasons why in the past there was no real dialogue between the business and the IT sides was a lack of good data, explains Business Engine's Mr O'Neil.

The firm, which has its roots in the defence industry, also helps companies with a technique at the cutting edge of technology management: balancing IT projects in the same way that many investors optimise their portfolios. Like different types of financial assets, IT projects can be classified according to risk and potential returns, allowing firms to pick a selection that fits their particular business strategy.

Peter Weill, a professor at MIT's Sloan School of Management, suggests that firms divide up their IT projects among four buckets representing different management objectives: cost reduction, better information, shared infrastructure and competitive advantage. Risk-averse and cost-conscious companies should have more projects in the first two buckets, whereas firms that put a premium on agility and are not afraid of failures should weight their portfolio in favour of the other two categories.

Who calls the shots?

All this fancy footwork, however, says Mr Weill, is not worth much without effective IT governance, by which he means rules that specify who makes the decisions and who is accountable. If his study of 265 companies in 23 countries is representative, most IT decisions – and not just those on geeky subjects such as picking the right IT infrastructure or architecture – are currently taken by technologists.

Some companies have already started to rebalance their governance. Merrill Lynch, for example, has put business people in charge of their technology portfolio. One of the things they have to do to get a project approved is to calculate its cost over five years, which they have a strong incentive to get right because these costs are charged back to a project's sponsors. They also have to reassess every quarter whether it is still viable. Financial markets can change very rapidly, so a project begun in 2000 to increase the capacity to process Nasdaq trades, for example, no longer makes much sense today.

Schneider National goes even further. It has an IT steering committee that acts like a venture-capital firm, screening all proposed IT projects and picking those with the best business plans. But the firm's in-house entrepreneurs do more than produce good ROI numbers. They also point out the necessary changes in business processes and organisation to ensure that employees are willing to use the new IT system. "People can undermine any technology," says Mr Lofgren.

Given the chill in the industry, it is no wonder that companies everywhere are rationalising their existing IT infrastructure and keeping purse strings tight. General Motors, for instance, has reduced the number of its computer systems from 3,000 to 1,300 by consolidating applications and servers. Merrill Lynch has cut its annual IT budget from $3 billion to $2 billion, mostly through what UBS Warburg, another investment bank, calls "cold technologies" – the sort that do not create new revenues for IT firms and often actually reduce spending. One of these is Linux, the free open-source operating system. Another is web services, which allow companies to integrate existing gear cheaply, thus giving new life to old, "legacy", systems such as mainframes.

No wonder, either, that companies spend their IT dollars differently from the way they used to. Software vendors, in particular, can no longer depend on quick multimillion-dollar deals, but must work hard to get much smaller contracts. Customers want bite-sized projects with quick returns, and increasingly pay up only if they are successful.

The danger of this new buying pattern is that companies may miss out on important long-term "architectural" investments, says John Hagel, a noted IT consultant. If vendors want IT spending to pick up again, they will have to concentrate more of their efforts on selling to business people rather than technologists. Yet many firms are "still stuck in the old world", he complains.

Luckily for IT companies, there is one customer that is spending more now than it did during the internet bubble: government. And that is only one of the reasons why the IT industry is becoming more involved in Washington, DC.

Regulating rebels

Despite its libertarian ideology, the IT industry is becoming increasingly involved in the machinery of government

EVEN AMONG THE GENERALLY LIBERTARIAN Silicon Valley crowd, T.J. Rodgers stands out. In early 2000, when everybody else was piling into the next red-hot initial public offering, the chief executive of Cypress Semiconductor, a chipmaker, declared that it would not be appropriate for the high-tech industry to normalise its relations with government. "The political scene in Washington is antithetical to the core values that drive our success in the international marketplace and risks converting entrepreneurs into statist businessmen," he wrote in a manifesto published by the Cato Institute, a think-tank.

A laudable sentiment, but in real life things are more complicated than that. In a sense, Silicon Valley is a creation of government. Without all the money from the military establishment, the region around San Jose would probably still be covered with fruit orchards. In any case, Mr Rogers's worst fears appear to be coming true. America's technology industry is becoming more and more intertwined with government. It has realised that the machinery of government in Washington can greatly influence its growth and profitability, and is becoming increasingly involved in lobbying. Conversely, the American government has become keenly aware of IT's crucial importance for the nation's well-being, heightened by the new emphasis on homeland security.

This should not come as a surprise, argues Debora Spar, a professor at Harvard Business School. "When technologies first emerge, there is a rush away from government and a surge of individualism. Over time, however, the rebels tend to return to the state," she writes in her book *Ruling the Waves* (Harcourt, 2001). And if the rebels become too powerful, the state tries to rein them in.

Take the development of the telegraph, in which government played an important role even though it was mainly driven by private firms. In the early days, the state protected the patents of Samuel Morse (who originally wanted government to fund and control the technology he had invented in 1835 because "this mode of instantaneous communication must inevitably become an instrument of immense power"). Later, the US Congress passed several laws regulating Western Union, the company that had monopolised telegraphy.

The reason public rules usually find their way into a technology, Ms Spar argues, is because government can protect property rights and restore order. But it also happens when a technology becomes widely used. "We cannot say the internet will have a huge influence on everyday life, and also say 'Hey Washington, keep out of it'," says Les Vadasz, a senior manager at Intel (who retired in 2003).

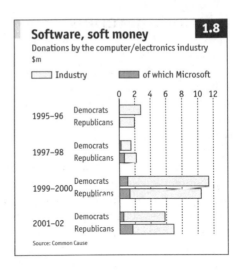

Software, soft money 1.8
Donations by the computer/electronics industry
$m

☐ Industry   ▉ of which Microsoft

0   2   4   6   8   10   12

1995–96 Democrats / Republicans

1997–98 Democrats / Republicans

1999–2000 Democrats / Republicans

2001–02 Democrats / Republicans

Source: Common Cause

Discovering a conscience

The chipmaker never had any ideological qualms about co-operating with government. In that sense, it has always been a mature company. Intel benefited from government money in the 1980s when it came under competitive pressure from Japanese manufacturers. Other Silicon Valley firms, too, owe much to the state. Oracle, for instance, grew out of a consulting job for the CIA, and the taxpayer stumps up for over one-fifth of its orders.

The Valley as a whole, however, did not develop a political conscience until 1996, when it successfully campaigned against a California ballot initiative that would have made sharcholder lawsuits much easier. This alerted the region's leaders to the need to get more involved to defend their interests, leading to the creation of such groups as Tech-Net, a lobbying and fund-raising organisation.

This environment also provided fertile ground for having a go at Microsoft. The antitrust case against the company might never have been brought without its competitors stirring up the trustbusters. The trial itself led to mass lobbying by both sides, as well as a rise in campaign-finance contributions. In fact, Microsoft has become one of the biggest donors to the Republican Party (see Chart 1.8).

Now that the IT industry's growth has slowed, the issues have changed. High-tech has discovered the Washington pork-barrel, a development that the Cato Institute calls a "Digital New Deal". At the top of the wish list is much more widespread high-speed internet access, or broadband. The number of connections has risen faster than expected, but companies such as Intel and Microsoft still think government should do something to push broadband, which would increase demand for high-tech goods and services. Security and privacy issues too are a high priority.

Yet the sector's most important political battle will be over property rights. Two conferences in 2003, one in Silicon Valley and one near it, highlighted the issues. "The law and technology of digital-rights management", was the theme of the event at the University of California at Berkeley. "Spectrum policy: property or commons?", asked the organisers at Stanford University.

To be sure, in technical terms intellectual property and radio spectrum are altogether different issues, but they pose similar policy challenges. In both cases technology is unsettling the status quo: the balance in copyright and the bureaucratic allocation of frequencies. And in both cases the main question now is how to organise markets to maximise innovation and investment.

The corporate interests battling it out in Washington naturally take a less lofty view. Hollywood wants nothing less than anti-piracy systems built into every electronic device, and is threatening to use its formidable lobbying power to get the legislation through if the IT industry does not comply voluntarily. Silicon Valley, worried that it will have to include government-dictated technology in its gear, has launched a huge lobbying campaign.

Battle for survival

The outcome of this battle, many industry experts argue, will determine to a large extent how fast the IT industry will grow. Without a balanced solution, online media and other advanced broadband services are unlikely ever to take off. If its wares are not sufficiently protected online, Hollywood will not make them available. And if electronic devices are put into a technological straitjacket, consumers will not use them.

The dispute surrounding the allocation of frequencies, triggered by the success of wireless internet access, known as Wi-Fi, might turn out to be even more important. The incumbents that currently own wide swathes of the radio spectrum, such as TV stations and cellular carriers, will fight tooth and nail to defend the status quo. Silicon Valley, for its part, is pushing for more of the spectrum to become a commons for everybody to use, which is what happens with Wi-Fi.

All this may sound like Washington business as usual, but the American government has definitely acquired a new interest in high-tech. Even before the terrorist attacks on September 11th 2001, it had identified the internet as part of the nation's critical infrastructure in need of better protection. Now IT is playing a central role in the war on terrorism and homeland security, both as a means to gather information and to improve the connections between different government agencies.

Shortly after September 11th, Congress passed the Patriot Act, which gives law enforcers new surveillance powers, such as monitoring internet traffic without a court order. Later, the Bush administration launched its Total Information Awareness (TIA) initiative, a highly controversial system developed by the Pentagon to sift through the electronic transactions of millions of people to spot suspicious activity.

All this makes government a key customer rather than just a big buyer. Most large enterprise software firms have launched "homeland security" initiatives in the hope of providing the federal government with technology to integrate its disparate databases to allow it to identify possible terrorists.

Vendors have also made their top engineers and researchers available as advisers, and are adapting their plans to reflect the fact that security has become the main priority. Some in Silicon Valley now liken the climate to that of the late 1970s, when government and military contractors employed more than 20% of the region's workforce.

Will the IT industry ever become as intertwined with government as, say, the car or media sectors? Nobody knows; but if it does, says Google's Eric Schmidt, high-tech will lose its innovative spark and, just like other sectors, turn to rent-seeking.

Déjà vu all over again

If history is any guide, the IT industry's future will be about services and customer power

Y ou WOULD EXPECT ERIC SCHMIDT, one of Silicon Valley's leading lights, to have an oversized inner geek. But these days, he sounds more like a closet historian. He enjoys talking, for instance, about how America's transcontinental railroad in the 1860s was built on debt, a bubble and scandals. Another favourite topic is the laying of the first transatlantic cable in that period, a seemingly impossible mission.

To Mr Schmidt, reading and thinking about history is a kind of redemption, for himself as well as for the high-tech industry: "We believed that the bubble would never end. We were wound up in a state of hubris." But of course, he says, it was déjà vu all over again: "People in high-tech didn't take any history classes."

If history is any guide, what does it tell us about the way the IT industry will evolve? As a technological revolution matures, its centre of gravity tends to shift from products to services. In railways, for instance, equipment-makers and train operators struggled, but lots of money was made by firms that used the railway infrastructure to offer new kinds of services, explains Brad DeLong at the University of California at Berkeley. One example was Sears, Roebuck & Co, which brought city goods to rural areas by mail order, offering a cheaper alternative to high-priced rural stores. In the same way, after the radio bubble, it was not the makers of the hardware that benefited most from the new medium, but broadcasters such as CBS.

A similar shift is bound to take place in the IT industry, predicts Geoffrey Moore of the Chasm Group. He says the sector's traditional business models are past their prime. Software firms, for instance, have made much of their money from shrink-wrapped products and platforms such as operating systems and databases. Increasingly, selling services of all kinds would be a better business to be in.

But it is not just IT firms that are becoming service providers, writes David Moschella in his recent book, *Customer-Driven IT* (Harvard Business School Press, 2003). IT customers themselves are moving in this direction. Instead of buying computer systems to become more efficient in their own business, he says, they will increasingly be using IT to create services for their clients, in effect becoming part of the sector's supply chain.

Obvious examples are internet companies such as Amazon, E*Trade and eBay. But it is increasingly useful, says Mr Moschella, to think of more traditional firms such as

Best bets   1.9
% share of the IT industry's profit, by sector

		1999	2000	2005	Trend
Business consultancy		10	13	17	▲
Services & software		30	29	41	▲
Hardware:	Server/storage	18	20	12	▼
	Clients	13	9	6	▼
Technology		29	29	24	▼

Source: IBM

banks, insurance companies and publishers as if they were a new kind of IT supplier: "All of the above are now in the business of systematically creating IT value for IT users, much as software and services companies have done in the past." From the user's point of view, there is not much difference between an online banking site and Microsoft's Hotmail service.

Being on top of the value chain, argues Mr Moschella, customers will increasingly be the industry's driving force. He urges them to band together and jointly develop new applications, platforms and standards in the same way that the financial industry has created credit cards and networks of ATMs. Such efforts could turn into "industry operating systems", a term coined by Tom Berquist, an analyst with Smith Barney: huge IT hubs that will take over many of the functions common to firms in a particular sector.

All of this suggests that IT customers will capture more of the sector's rent. But even if things play out differently, the balance of power is likely to shift away from vendors and in favour of buyers. Having learnt the painful lessons of over-investment during the boom, they will no longer allow themselves to be locked into proprietary technology.

Nimbler than airlines

So will IT firms end up, in Mr Schmidt's worst-case scenario, "like today's airlines", which always seem to be in or close to Chapter 11? Fortunately for shareholders, they probably won't, at least in the foreseeable future – for the simple reason that they will make active efforts to prevent such a calamity. In fact, vendors are already changing their business models, mostly by moving up the technology stack. Sun, which made a killing during the dotcom boom by selling high-end servers, is trying to become more of a software firm and a builder of power plants for computing utilities. And much of Microsoft's .NET effort is about software as a service.

Yet it is IBM that is betting most on the prediction that the IT industry will follow historic patterns of evolution. Big Blue expects profits to migrate to software and services (see Chart 1.9), and is managing its product portfolio accordingly. For example, it has sold its hard-drive business and acquired the consulting arm of PricewaterhouseCoopers, an accountancy firm. Slowly but surely, IBM is morphing from a technology vendor with a strong IT-services arm into a business consulting firm that also sells software and hardware.

Bigger and better

The industry has also already begun to consolidate in response to the shifting balance of power. The merger of Compaq and HP looks much more prescient today than when it was announced in September 2001. Future corporate marriages will not necessarily be that huge, but there will be many of them. Oracle's Mr Ellison says there are at least 1,000 Silicon Valley companies that need to go bankrupt.

Such a mass exodus, again, would not be without historical precedent. Most industries have seen shake-outs when they grow up, says Steven Klepper, an economic historian at Carnegie Mellon University. In America's car industry, for instance, the number of producers peaked at 274 in 1909. By 1918, it had dropped to 121. By 1955, only seven were left.

The car industry is also instructive in that much of its production has been outsourced to suppliers. Similarly, predict George Gilbert and Rahul Sood, two software-industry analysts, software firms will now develop something they never had before: a supply chain. In a way, open-source is an early incarnation of this: a veritable army of volunteer programmers contribute patches to software such as Linux. In future, Messrs Gilbert and Sood predict, a large part of software development will be "outshored" to countries such as India and China, which are already generating much code (and not just the easy stuff). This will mean that big software vendors will become more like aggregators. At least one of them, SAP, is aiming at exactly that. It wants suppliers to develop applications, so-called xapps, and assemble them along with its own components into software suites.

But perhaps the best news for the industry is that there are still plenty of opportunities in the new world of IT. "If we go with the market, help our customers to realise the business value of IT, then we can be a good business," says IBM's Mr Wladawsky-Berger. For a start, all that experimentation during the dotcom boom actually produced some useful results. Things tried during a technological bubble tend to make a comeback. The first transatlantic cable, for example, was a disaster, but it prompted others to try again.

Most business-to-business marketplaces failed dismally, because these start-ups thought technology would quickly overthrow existing power structures, explains Mr Moore. But these firms got one thing right: there are lots of assets trapped in inefficient supply chains. Established institutions are now pragmatically adopting these technologies, for instance in the form of private exchanges controlled by buyers.

And there still remain many more new things to try out, which is where IT arguably differs most from previous revolutions. Whether railways, cars or even electricity, all are relatively limited technologies compared with IT, which in time is likely to embrace the whole of business and society.

Currently, wireless technologies are all the rage, although again nobody knows how much money will be in it for vendors and carriers. Optimists hope that surfers will soon be able to roam around freely and remain continuously connected to the internet. And small radio chips called RFID tags will make it possible to track everything and anything, promising to make supply chains much more efficient. But even a new killer application is unlikely to bring back the good old times. "After a crash, much of the glamour of the new technology is gone," writes Brian Arthur, an economist at the Santa Fe Institute. The years after the British railway mania, for instance, were "years of build-out rather than novelty, years of confidence and steady growth, years of orderliness."

This kind of "new normal", in the words of Accenture, another IT consultancy, may be hard to swallow for a sector that has always prided itself on being different. But for its customers, a more mature IT industry is a very good thing: just as the best technology is invisible, the best IT industry is one that has completely melted into the mainstream. Thriving on exponentials was certainly fun. But even paradise can get boring after a while.

POSTSCRIPT

Since this section was published in 2003, the IT industry has continued to mature and consolidate. There has been a series of mergers and takeovers, including Oracle's purchase of PeopleSoft, the tie-up between Veritas and Symantec, Adobe's purchase of Macromedia, and IBM's sale of its PC division to Lenovo, a Chinese firm. Larry Ellison of Oracle has acted as the cheerleader for the idea that the industry will never re-capture the glories of its youth. "It's not coming back," he said in an interview in 2003. "The industry is maturing. The valley will never be what it was."

The material on pages 4–40 first appeared in a survey in *The Economist* in May 2003.

2
SECURING THE CLOUD

Securing the cloud

Digital security, once the province of geeks, is now everyone's concern. But there is much more to the problem – or the solution – than mere technology

W HEN THE WORLD'S RICHEST MAN decides it is time for his company to change direction, it is worth asking why. Only rarely does Bill Gates send an e-mail memo to the thousands of employees at Microsoft, the world's largest software company, of which he is chairman. He famously sent such a memo in December 1995, in which he announced that Microsoft had to become "hardcore" about the internet. In January 2002 Mr Gates sent another round-robin. Its subject? The importance of computer security.

Until recently, most people were either unaware of computer security or regarded it as unimportant. That used to be broadly true, except in a few specialised areas – such as banking, aerospace and military applications – that rely on computers and networks being hard to break into and not going wrong. But now consumers, companies and governments around the world are sitting up and taking notice. Why?

The obvious answer seems to be that the 2001 terrorist attacks in America heightened awareness of security in all its forms. But the deeper reason is that a long-term cultural shift is under way. Digital security has been growing in importance for years as more and more aspects of business and personal life have come to depend on computers. Computing, in short, is in the midst of a transition from an optional tool to a ubiquitous utility. And people expect utilities to be reliable. One definition of a utility, indeed, is a service that is so reliable that people notice it only when it does not work. Telephone service (on fixed lines, at least), electricity, gas and water supplies all meet this definition. Computing clearly does not, at least not yet.

One of the many prerequisites for computing to become a utility is adequate security. It is dangerous to entrust your company, your personal information or indeed your life to a system that is full of security holes. As a result, the problem of securing computers and networks, which used to matter only to a handful of system administrators, has become of far more widespread concern.

Computers are increasingly relied upon; they are also increasingly connected to each other, thanks to the internet. Linking millions of computers together in a single, cloud-like global network brings great benefits of cost and convenience. Dotcoms may have come and gone, but e-mail has become a vital business tool for many people and an important social tool for an even larger group. Being able to access your e-mail from any web browser on earth is tremendously useful and liberating, as both business travellers and backpacking tourists will attest. Corporate billing, payroll and inventory-tracking systems are delivered as services accessible through web browsers. Online shop fronts make it fast and convenient to buy products from the other side of the world.

The price of openness

The flip side of easy connectivity and remote access, however, is the heightened risk of a security breach. Bruce Schneier, a security expert, points out that when you open a shop on the street, both customers and shoplifters can enter. "You can't have one without the other," he says. "It's the same on the internet." And as music, movies, tax returns, photographs and phone calls now routinely whizz around in digital form, the shift from traditional to digital formats has reached a critical point, says Whitfield Diffie, a security guru at Sun Microsystems: "We can no longer continue this migration without basic security."

The September 11th attacks, then, reinforced an existing trend. Government officials, led by Richard Clarke, America's cyber-security tsar, gave warning of the possibility that terrorists might mount an "electronic Pearl Harbour" attack, breaking into the systems that control critical telecommunications, electricity and utility infrastructure, and paralysing America from afar with a few clicks of a mouse. Most security experts are sceptical, but after spending years trying to get people to take security seriously, they are willing to play along. Scott Charney, a former chief of computer crime at the Department of Justice and now Microsoft's chief security strategist, says Mr Clarke's scare-mongering is "not always helpful, but he has raised awareness".

The terrorist attacks certainly prompted companies to acknowledge their dependence on (and the vulnerability of) their networks, and emphasised the importance of disaster-recovery and back-up systems. A survey of information-technology managers and chief information officers, carried out by Morgan Stanley after the attacks, found that security software had jumped from fifth priority or lower to become their first priority. "It's moved up to the top of the list," says Tony Scott, chief technology officer at General Motors. "It's on everybody's radar now."

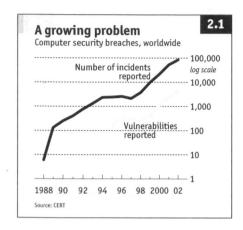

A growing problem `2.1`
Computer security breaches, worldwide

Number of incidents reported

Vulnerabilities reported

log scale

100,000
10,000
1,000
100
10
1

1988 90 92 94 96 98 2000 02

Source: CERT

The growing emphasis on security has been driven by a combination of factors, and has shown up in a variety of ways. Chris Byrnes, an analyst at Meta Group, a consultancy, notes that the proportion of his firm's clients (mostly large multinational companies) with dedicated computer-security teams rose from 20% to 40% between 2000 and 2002. Previously, he says, it was financial-services firms that were most serious about security, but now firms in manufacturing, retailing and other areas are following suit.

One important factor is regulation. Mr Byrnes points to the change made to American audit standards in 1999, requiring companies to ensure that information used to prepare public accounts is adequately secured. This has been widely interpreted, with the backing of the White House's critical-infrastructure assurance office, to mean that a company's entire network must be secure.

Similarly, the April 2003 deadline for protecting patients' medical information under the Health Insurance Portability and Accountability Act (HIPAA) prompted health-care providers, pharmaceutical companies and insurers to re-evaluate and overhaul the security of their computers and networks. In one case, Eli Lilly, a drugmaker, was accused of violating its own online privacy policy after it accidentally revealed the e-mail addresses of 669 patients who were taking Prozac, an anti-depressant. The company settled out of court with America's Federal Trade Commission and agreed to improve its security procedures. But once HIPAA's privacy regulations come into force, companies that fail to meet regulatory standards will face stiff financial penalties. The same sort of thing is happening in financial services, where security is being beefed up prior to the introduction of the Basel II bank-capital regulations.

The growth of high-profile security breaches has also underlined the need to improve security. The number of incidents reported to Carnegie Mellon's computer emergency response team (CERT), including virus outbreaks and unauthorised system intrusions, has shot up in recent years (see Chart 2.1) as the internet has grown. The "Love Bug", a virus that spreads by e-mailing copies of itself to everyone in an infected computer's address book, was front-page news when it struck in May 2000. Many companies, and even Britain's Parliament, shut down their mail servers to prevent it from spreading.

There have been a number of increasingly potent viruses since then, including Sircam,

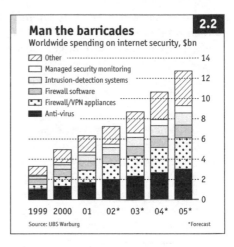

Man the barricades — 2.2
Worldwide spending on internet security, $bn

- Other
- Managed security monitoring
- Intrusion-detection systems
- Firewall software
- Firewall/VPN appliances
- Anti-virus

1999 2000 01 02* 03* 04* 05*

Source: UBS Warburg   *Forecast

Code Red and Nimda, all of which affected hundreds of thousands of machines. Viruses are merely one of the more visible kinds of security problem, but given the disruption they can cause, and the widespread media coverage they generate, such outbreaks prompt people to take security more seriously.

Fear, sex and coffee

Spending on security technology grew by 28% in 2001 compared with the year before, according to Jordan Klein, an analyst at UBS Warburg. Mr Klein predicted that spending would continue to grow strongly, from around $6 billion in 2001 to $13 billion in 2005 (see Chart 2.2). A survey carried out by Meta Group in August 2002 found that although only 24% of firms had increased their technology budgets in 2002, 73% had increased their spending on security, so security spending is growing at the expense of other technology spending. This makes it a rare bright spot amid the gloom in the technology industry.

Steven Hofmeyr of Company 51, a security start-up based in Silicon Valley, says his company is pushing at a wide-open door: there is no need to convince anyone of the need for security technology. Indeed, Nick Sturiale of Sevin Rosen, a venture-capital fund, suggests that security is already an overcrowded and overfunded sector. "Security is now the Pavlovian word that draws the drool from VCs' mouths," he says. Security vendors are really selling fear, he says, and fear and sex are "the two great sales pitches that make people buy irrationally".

So, a bonanza for security-technology firms? Not necessarily. The sudden interest in security does not always translate into support from senior management and larger budgets. A recent report from Vista Research, a consultancy, predicts that: "While the need to protect digital assets is well established, companies will pay lip service to the need to invest in this area and then largely drag their feet when it comes to capital spending on security."

Even where security spending is increasing, it is from a very low base. Meta Group's survey found that most companies spend less than 3% of their technology budgets on security. Technology budgets, in turn, are typically set at around 3% of revenues. Since 3% of 3% is 0.09%, most firms spend more on coffee than on computer security, according to a popular industry statistic. The purse strings loosen only when companies suffer a serious security breach themselves, see one of their rivals come under attack or are told by auditors that lax security could mean they are compromising due diligence.

Jobs on plates

Mr Byrnes notes another factor that is impeding growth of the security market: a shortage of senior specialists. For much of 2002, he says, "There was more security budget than ability to spend it." John Schwarz, president of Symantec, a security firm, put the number of unfilled security jobs at 75,000 in America alone.

Given the interest in security, established technology firms, which have seen revenues plunge as firms slash technology spending in other areas, are understandably keen to jump on the bandwagon alongside specialist security vendors. Sun's advertisements boast: "We make the net secure." Oracle, the world's second-largest software firm, has launched a high-profile campaign trumpeting (to guffaws from security experts) that its database software is "unbreakable". Whether or not this is true, Oracle clearly regards security as a convenient stick with which to bash its larger arch-rival, Microsoft, whose products are notoriously insecure – hence Mr Gates's memo.

It suits vendors to present security as a technological problem that can be easily fixed with more technology – preferably theirs. But expecting fancy technology alone to solve the problem is just one of three dangerous misconceptions about digital security. Improving security means implementing appropriate policies, removing perverse incentives and managing risks, not just buying clever hardware and software. There are no quick fixes. Digital security depends as much – if not more – on human cultural factors as it does on technology. Implementing security is a management as well as a technical problem. Technology is necessary, but not sufficient.

A second, related misperception is that security can be left to the specialists in the systems department. It cannot. It requires the co-operation and support of senior management. Deciding which assets need the most protection, and determining the appropriate balance between cost and risk, are strategic decisions that only senior management should make. Furthermore, security almost inevitably involves inconvenience. Without a clear signal from upstairs, users will tend to regard security measures as nuisances that prevent them from doing their jobs, and find ways to get around them.

Unfortunately, says Mr Charney, senior executives often find computer security too complex. "Fire they understand," he says, because they have direct personal experience of it and know that you have to buy insurance and install sensors and sprinklers. Computer security is different. Senior executives do not understand the threats or the technologies. "It seems magical to them," says Mr Charney. Worse, it's a moving target, making budgeting difficult.

A third common misperception concerns the nature of the threat. Even senior managers who are aware of the problem tend to worry about the wrong things, such as virus outbreaks and malicious hackers. They overlook the bigger problems associated with internal security, disgruntled ex-employees, network links to supposedly trustworthy customers and suppliers, theft of laptop or handheld computers and insecure wireless access points set up by employees. That is not surprising: viruses and hackers tend to get a lot of publicity, whereas internal security breaches are hushed up and the threats associated with new technologies are often overlooked. But it sets the wrong priorities.

Detective stories

A final, minor, misperception is that computer security is terribly boring. In fact, it turns out to be one of the more interesting aspects of the technology industry. The war stories told by security consultants and computer-crime specialists are far more riveting than discussion of the pros and cons of customer-relationship management systems. So there really is no excuse for avoiding the subject.

Anyone who has not done so already should take an interest in computer security. Unfortunately there is no single right answer to the problem. What is appropriate for a bank, for example, would be overkill for a small company. Technology is merely part of the answer, but it has an important role to play.

Tools of the trade

How a box of technological tricks can improve (but not guarantee) your security

ASK A NON-SPECIALIST about computer security, and he will probably mention viruses and attacks by malicious hackers, if only because they are so much more visible than other security problems. Take viruses first. Like their biological counterparts, computer viruses are nasty strings of code that exploit their hosts to replicate themselves and cause trouble. Until a few years ago, viruses merely infected files on a single computer. Eventually, an infected file would be moved, typically on a floppy disk, to another machine where the virus could spread. Modern viruses, however, are far more insidious, because they can jump from one computer to another across the internet, most often by e-mail. (Since self-propagating programs are technically known as worms, such viruses are sometimes called worm-virus hybrids.)

High-profile examples include Sircam, which struck in July 2001 and generated much comment because as well as e-mailing copies of itself to everyone in an infected PC's address book, like previous viruses, it also enclosed random documents from the infected machine's hard disk with each message. Users thus unwittingly sent half-finished screen-plays, unsent letters and private diary jottings to their friends, some-times with embarrassing results. Code Red, which also struck that month, was a worm that exploited a security vulnerability in Microsoft's web-server software to spread from one server to another. Infected servers were programmed to flood the White House website with traffic for a week.

Patching it up

The weakness that Code Red exploited had been discovered in June, and Microsoft had issued a software "patch" to correct it. But software patches are issued all the time, and keeping up with new patches and deciding which to install is more than many system administrators can manage. Within a week of Code Red's appearance, 300,000 computers were infected with it. Sometimes it defaced infected web servers with the message "Hacked by Chinese!", which suggested a political moti-vation, but the identities and motives of virus writers can rarely be determined for sure. Similarly, Nimda, a particularly vigorous virus/worm which struck on September 18th 2001, was initially assumed to have some connection with the previous week's terrorist attacks, though this now seems unlikely.

Viruses are extremely widespread, which is more than can be said for meaningful statistics about them. The annual survey carried out by the Computer Security Institute (CSI) in San Francisco, in conjunction with the Federal Bureau of Investigation's local computer-intrusion squad, is generally regarded as one of the more authoritative sources of information about computer security. According to a CSI/FBI report published in April 2002, 85% of respondents (mainly large American companies and government agencies) encountered computer viruses during 2001 (see Chart 2.3). Quantifying the damage done by viruses, however, is extremely difficult. Certainly, cutting off e-mail or internet connections can seriously hamper a company's ability to do business. In severe cases every single computer in an office or school may need to be disinfected, which can take days.

Yet assigning costs to outbreaks is guesswork at best. Computer Economics, a consultancy, put the worldwide costs imposed by viruses in 2001 at $13.2 billion. But few outside the marketing departments of anti-virus-software vendors take such figures seriously. Critics point out that if most companies are themselves unable to quantify the cost of cleaning up viruses in their systems, it is hard to see how anyone else can. Far easier to quantify is the surge in sales of anti-virus software that follows each outbreak. Following the Code Red and Nimda strikes, for example, anti-virus sales at Symantec, a leading security-software firm, in the last quarter of 2001 were 53% up on a year earlier.

Anti-virus software works by scanning files, e-mail messages and network traffic for the distinguishing characteristics, or "signatures", of known viruses. There is no general way to distinguish a virus from a non-malicious piece of code. Both are, after all, just computer programs, and whether a particular program is malicious or not is often a matter of opinion. So it is only after a virus has infected its first victims and has started to spread that its signature can be determined by human analysts, and that other machines can be inoculated against it by having their database of signatures updated. Inevitably, the result is an arms race between the mysterious folk who write viruses (largely for fun, it seems, and to win the kudos of their peers) and makers of anti-virus software. Some viruses, including one called Klez, even attempt to disable anti-virus software on machines they infect, or spread by posing as anti-virus updates.

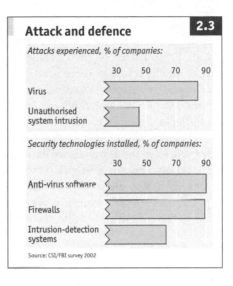

Attack and defence 2.3

Attacks experienced, % of companies:

	30	50	70	90
Virus				
Unauthorised system intrusion				

Security technologies installed, % of companies:

	30	50	70	90
Anti-virus software				
Firewalls				
Intrusion-detection systems				

Source: CSI/FBI survey 2002

Viruses are a nuisance, but the coverage they receive is disproportionate to the danger they pose. Some vendors of anti-virus software, particularly the smaller ones, fuel the hysteria by sending out jargon-filled warnings by e-mail at every opportunity. From a technical point of view, protecting a computer or network against viruses is tedious but relatively simple, however: it involves installing anti-virus software on individual machines and keeping it up to date. Virus-scanning software that sits on mail servers and scans e-mail messages before they are delivered can provide an extra defensive layer.

Dealing with intrusions by malicious hackers is an altogether more complex problem. (The word "hacker" merely means a clever programmer, but is commonly applied to those who use their skills to malicious ends.) Computers are such complex systems that there are endless ways for unauthorised users to attempt to gain access. Attackers very often use the same security flaws that worms and viruses exploit; such worms and viruses can be seen as an automated form of malicious hacking.

Having gained access to a machine, an attacker can deface web pages (if the machine is a web server), copy information (if the machine stores user information, financial data or other documents), use the machine as a base from which to attack other machines, or install "Trojan horse" software to provide easy access in future or to enable the machine to be remotely controlled over the internet. Savvy attackers cover their tracks using special software known as a "root kit", which conceals the evidence of their activities and makes unauthorised use difficult to detect.

As with viruses, meaningful figures for unauthorised intrusions are hard to find. Many attacks go unnoticed or unreported. But the CSI/FBI survey gives some flavour of the scale of the problem. Of the 503 large companies and government agencies that participated in the survey,

40% detected system intrusions during 2001, and 20% reported theft of proprietary information. Of the companies that were attacked, 70% reported vandalism of their websites. But it is dangerous to lump all attacks together. Just as there is a difference between a graffiti-spraying youth and a criminal mastermind, there is a world of difference between vandalising a web page and large-scale financial fraud or theft of intellectual property.

The principal tool for keeping unwanted intruders out of computers or networks is the firewall. As its name suggests, a firewall is a device that sits between one network (typically the internet) and another (such as a closed corporate network), enforcing a set of rules about what can travel to and fro. For example, web pages might be allowed inside the firewall, but files might not be allowed to go outside.

Walls have ears

Firewalls are no panacea, however, and may give users a false sense of security. To be effective, they must be properly configured, and must be regularly updated as new threats and vulnerabilities are discovered. "What kind of firewall you have matters far less than how you configure it," says Bill Murray of TruSecure, a security consultancy. There are dozens of competing firewall products on the market, but most of them come in two main forms: as software, which can be installed on a machine to regulate traffic, and as hardware, in the form of appliances that plug in between two networks and regulate the flow of traffic between them.

The leader in the field is Check Point Software of Ramat Gan, Israel. In 1998, says Jerry Ungerman, Check Point's president, people thought the firewall market was almost saturated, because most firms had one, but the market has continued to grow. The notion that each company simply needs one firewall, between its internal network and the internet, is now outmoded. Companies often have many separate links to the internet, want to wall off parts of their internal networks from each other, or choose to install firewall software on every server. Some of Check Point's clients, says Mr Ungerman, have over 1,000 firewalls installed. The advent of fixed broadband connections means that home users, who often leave their computers switched on around the clock, now need firewalls too if they are to protect their machines from intruders. Even mobile phones and hand-held computers, he predicts, will have firewalls built into them.

Firewalls have their uses, but there are many kinds of attacks they cannot prevent. An attacker may be able to bypass the firewall, or exploit a vulnerability by sending traffic that the firewall regards as legitimate. Many attacks involve sending artfully formulated requests to web servers, causing them to do things that would not normally be allowed, says Geoff Davies of i-Sec, a British security consultancy. To show how easily this can be done, he types a string of database commands into the search field of an online travel agent, and instead of a table of flight departures and arrivals, the website comes up with a table of information about its users. (Mr Davies carried out this demonstration, called an "SQL insertion" attack, on a dummy server specially set up for the purpose, but it is a widespread vulnerability on real websites.) To a firewall, such an attack may look just like a legitimate use of the web server.

Halt! Who goes there?

An alternative is the "intrusion-detection system" (IDS), which monitors patterns of behaviour on a network or an individual computer and sounds an alarm if something looks fishy. Some kinds of detection systems monitor network traffic, looking for unusual activity, such as messages passing to and from a Trojan horse on the network; others sit on computers, looking for unusual patterns of access, such as attempts to retrieve password files.

Compared with anti-virus software and firewalls, detection is a relatively immature technology, and many people believe it is more trouble than it is worth. The difficulty is tuning an IDS correctly, so that it spots mischievous behaviour reliably without sounding too many false alarms. An IDS may end up like the boy who cried wolf – when a genuine attack occurs after too many false alarms, nobody pays any attention. And even when it is properly tuned, people may not know how to stop the problem when an IDS sounds the alarm. All too often the response is, "We just got hacked – what do we do?", says Chris King of Meta Group.

Other tools in the security toolbox include encryption, the mathematical scrambling of data so that only the intended recipient can read them, and the related technique of cryptographic authentication to verify that people are who they claim they are. These tools can be integrated into an e-mail system, for example, using encryption to ensure that messages cannot be read in transit, and authentication to ensure that each message really did come from its apparent sender. The same techniques can also be used to send information (such as credit-card details) to and from websites securely.

Another popular use of encryption and authentication is the "virtual private network" (VPN), which allows authenticated users to establish secure communications channels over the internet to a closed network. VPNs are widely used to knit a company's networks in different parts of the world together securely across the internet, and to allow travelling employees to gain secure access to the company network from wherever they are.

There is still plenty of room for innovation in security technology, and there are dozens of start-ups working in the field. Company 51 of San Mateo, California, has devised an "intrusion-prevention system", based on the workings of the human immune system. When an attack is detected, the attacker is promptly disconnected. Cenzic, also based in Silicon Valley, has devised a novel approach to security testing called "fault injection". Greg Hoglund, the company's co-founder, says most testing of security software is akin to testing a car by driving it on a straight, flat road. Just as cars are crash-tested, Cenzic's software, called Hailstorm, stress-tests software by bombarding it with attacks.

Blame it on the bugs

A typical network, then, is secured using a multilayered combination of security technologies. But these fancy measures merely treat the effects of poor security. A parallel effort is being made to deal with one of its main causes: badly written software. According to @Stake, a security consultancy based in Cambridge, Massachusetts, 70% of security defects are due to flaws in software design. Code Red, for example, exploited a "bug", or coding error, in the way Microsoft's web-server software handles non-Roman characters. Buggy software tends to be insecure. So by taking a firmer stand against bugs and making their programs more reliable, software firms can also improve security.

Microsoft is now making a particular effort to improve its reputation for shoddy security. New bugs and vulnerabilities in its products are found weekly. This does not necessarily mean that Microsoft's software is particularly badly written, but has much to do with its ubiquity. Microsoft has a monopoly in desktop operating systems, after all, and a near-monopoly in web browsers and office productivity software. Find a hole in Internet Explorer, Microsoft's web browser, for example, and you are capable of attacking the vast majority of the world's PCs. Find a hole in Netscape's rival web browser, which is far less widely used, and you will be able to attack fewer than 10% of them.

Now that the threat to Microsoft of dismemberment by America's

Department of Justice has receded, the company's poor reputation in security looks like its single biggest problem. In 2001, following the Code Red and Nimda outbreaks, both of which exploited security flaws in Microsoft products, John Pescatore at Gartner, an influential consultancy firm, suggested that companies that wanted to avoid further security problems should stop using Microsoft's software. Bill Gates responded by issuing his company-wide memo in January 2002 on "trustworthy computing".

Microsoft is pulling out all the stops to reduce the number of security vulnerabilities in its products. "There was a sea change in the way our customers were thinking," says Pierre de Vries, Microsoft's director of advanced product development. The company, he says, realised that it had "a real problem" even before Mr Pescatore's report. In 2002, the 8,500 programmers in the company's Windows division were given new security training, after which they spent two months combing their code for potential vulnerabilities. Tools devised by the company's research division, called "Prefix" and "Prefast", are used to scan for possible problems. And when coding errors are found, they are not only fixed but an effort is now made to find out how they slipped through the net in the first place.

Microsoft has also tightened security in other ways. Its web-server software, for example, now arrives with most options switched off by default. Customers have to decide which options they want to use, and make a conscious choice to switch them on. This reduces their exposure to problems in parts of the software they were not using anyway. But some customers complained about having to work out which options they did and did not need, says Mr de Vries. One of them even asked for a button to turn everything on. The cost of improved security, it seems, is often a reduction in convenience.

This kind of thing goes against the grain for Microsoft. Traditionally, its products have had all the bells and whistles (such as the infamous talking paper clip) turned on by default, to make it more likely that users will discover and use new features. Microsoft is also renowned for encouraging users to upgrade for extra features. But priorities have changed. As Mr Gates wrote to his workforce, "When we face a choice between adding features and resolving security issues, we need to choose security."

Microsoft's policy of tight integration between its products, which both enhances ease of use and discourages the use of rival software-makers' products, also conflicts with the need for security. Because

Microsoft's programs are all linked, a flaw in one of them can be used to gain access to others. Many viruses, for example, exploit holes in Microsoft's mail or browser software to infect the underlying Windows operating system.

Many observers believe that Microsoft's new-found concern over security is mere window-dressing. The Windows operating system is the largest piece of software ever written, so implementing security retrospectively is a daunting task. Mary Ann Davidson, chief security officer at Oracle, contends that American federal agencies are "really angry" with Microsoft over the insecurity of its products. Oracle, whose flagship database software grew out of a consulting contract for the Central Intelligence Agency, has many black-suited "professional paranoids" among its customers, so the company has security awareness far more deeply ingrained in its corporate culture, she says. But what of Oracle's advertising claims that its own software is "unbreakable"? Perfect security is impossible, she concedes; the campaign "is about being fanatical about security".

Need to know

A key test of a company's commitment to security is the speed with which it responds to vulnerabilities. The difficulty, says Steve Lipner, Microsoft's director of security assurance, is that when a new vulnerability is discovered, customers want a patch immediately, but they also want the patch to be properly tested, which takes time. Furthermore, issuing a separate patch for every vulnerability makes life harder for systems administrators, so Microsoft now prefers to group several patches together. But that lays it open to the charge that it is not responding fast enough. Once a vulnerability has been announced, attackers will start trying to exploit it immediately. According to Mr Davies, some big websites get attacked as little as 40 minutes after the publication of a new vulnerability. But the patch may not be available for weeks.

Mr Lipner says he would prefer researchers who find flaws to report them to Microsoft, but not to publicise them until a patch is available. The trouble is that software-makers have little incentive to fix patches that nobody knows about, so many security researchers advocate making vulnerabilities public as soon as they are found. Admittedly, this alerts potential attackers, but they may already have known about them anyway. Proponents of this "full disclosure" approach argue that its benefits outweigh the risks. "Sunlight is the best disinfectant," says Mr Diffie at Sun Microsystems.

Will software-makers ever come up with products that are free of security vulnerabilities? It seems very unlikely, but even if they did, there would still be plenty of systems that remained unpatched or incorrectly configured, and thus vulnerable to attack. No matter how clever the technology, there is always scope for human error. Security is like a chain, and the weakest link is usually a human.

The weakest link

If only computer security did not have to involve people

THE STEREOTYPE OF THE MALICIOUS HACKER is a pale-skinned young man, hunched over a keyboard in a darkened room, who prefers the company of computers to that of people. But the most successful attackers are garrulous types who can talk their way into, and out of, almost any situation. In the words of Mr Schneier, a security guru, "Amateurs hack systems, professionals hack people."

Kevin Mitnick, perhaps the most notorious hacker of recent years, relied heavily on human vulnerabilities to get into the computer systems of American government agencies and technology companies including Fujitsu, Motorola and Sun Microsystems. Testifying before a Senate panel on government computer security in 2000, after spending nearly five years in jail, Mr Mitnick explained that:

> When I would try to get into these systems, the first line of attack would be what I call a social engineering attack, which really means trying to manipulate somebody over the phone through deception. I was so successful in that line of attack that I rarely had to go towards a technical attack. The human side of computer security is easily exploited and constantly overlooked. Companies spend millions of dollars on firewalls, encryption and secure access devices, and it's money wasted, because none of these measures address the weakest link in the security chain.

Human failings, in other words, can undermine even the cleverest security measures. In one survey, carried out by PentaSafe Security, two-thirds of commuters at London's Victoria Station were happy to reveal their computer password in return for a ballpoint pen. Another survey found that nearly half of British office workers used their own name, the name of a family member or that of a pet as their password. Other common failings include writing passwords down on sticky notes attached to the computer's monitor, or on whiteboards nearby; leaving machines logged on while out at lunch; and leaving laptop computers containing confidential information unsecured in public places.

Unless they avoid such elementary mistakes, a firm's own employees may pose the largest single risk to security. Not even technical staff who should know better are immune to social engineering. According to Meta Group, the most common way for intruders to gain access to company systems is not technical, but simply involves finding out

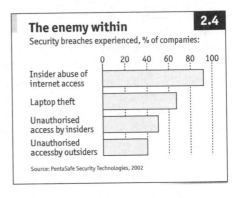

The enemy within 2.4

Security breaches experienced, % of companies:

Source: PentaSafe Security Technologies, 2002

the full name and username of an employee (easily deduced from an e-mail message), calling the help desk posing as that employee, and pretending to have forgotten the password.

Simple measures, such as encouraging employees to log out during lunch hours and to choose sensible passwords, can dramatically enhance security at very little cost. Passwords should be at least six and ideally eight characters long, and contain a mixture of numbers, letters and punctuation marks. Dictionary words and personal information should not be used as passwords. Users should have a different password on each system, and they should never reveal their passwords to anyone, including systems managers.

Yet a seminal paper published as long ago as 1979 by Ken Thomson and Robert Morris found that nearly a fifth of users chose passwords consisting of no more than three characters, and that a third used dictionary words. (Robert Morris, the chief scientist at America's National Computer Security Centre, was subsequently upstaged by his son, also called Robert, who released the first internet worm in 1988 and crashed thousands of computers. Ironically, the worm exploited badly chosen passwords.) But back in 1979, only a small fraction of a typical company's workforce used computers on a daily basis. Now that almost everybody uses them, the potential for trouble is much greater.

A few precautions also go a long way when it comes to stopping the spread of viruses. Many viruses travel inside e-mail messages, but require the user to double-click them in order to start propagating. So they pose as games, utilities, anti-virus updates or even as nude photographs of well-known tennis players. The curious user double-clicks, nothing seems to happen, and the user thinks no more about it, but the virus has started to spread. Educating users not to double-click on dubious attachments is a simple but effective counter-measure against viruses.

If correctly handled, a management-based, rather than a solely technology-based, approach to security can be highly cost-effective. The danger, says Peter Horst of TruSecure, is that: "People buy a hunk of shining technology, wipe their brow and say, 'Great, I've taken care of it,' when they might have been better off saving money and doing something simple in terms of policy and process." Probably the best example of how expensive, glamorous security technology can easily be undermined by poor procedures is biometric systems (see page 64).

A sensible and balanced approach, then, involves not only security technology but also a well-defined set of security policies which users understand and keep to. This approach is promoted by the Human Firewall Council, a group which argues that users themselves have an important role to play in maintaining security. Steve Kahan, its president, draws an analogy with neighbourhood-watch schemes. The idea, he says, is "to make security everyone's business", and to have a clear security policy that governs what is and is not allowed. That policy should then be implemented both by guiding the behaviour of users and by the appropriate configuration of firewalls, anti-virus software and so forth, in much the same way that a combination of neighbourly vigilance, alarms and door locks is used to combat burglars in the real world. But, says Mr Kahan, surveys show that half of all office workers never receive any security training at all.

One way to disseminate and enforce security policy is to add yet another layer of security software, as demonstrated by PentaSafe Security, one of the backers of the Human Firewall Council. Its software can ensure that users are familiar with a company's security policy by popping messages and quiz-like questions up on the screen when they log on. According to PentaSafe's figures, 73% of companies never require employees to re-read security policies after they begin their employment, and two-thirds of companies do not track whether their employees have read the policy in the first place.

David Spinks, European director of security at EDS, a computer-services firm, says all EDS employees have to take a regular on-screen test to ensure they understand the company's policy on passwords, viruses and network security. Choice of technology, he says, matters far less than managing both technology and users properly: "The key to having a firewall isn't the firewall, but how the policies are set, monitored, managed and kept up to date." Two companies can use exactly the same product, he notes, and one can be secure while the other is insecure. It is effective management that makes the difference.

The dismal science of security

But there are other, more subtle ways in which management and security interact. "More than anything else, information security is about work flow," says Ross Anderson of Cambridge University's Computer Laboratory. The way to improve security, he says, is to think about people and processes rather than to buy a shiny new box. Mr Anderson is one of a growing number of computer scientists who are applying ideas from economic theory to information security. Insecurity, he says, "is often due to perverse incentives, rather than to the lack of suitable technical protection mechanisms." The person or company best placed to protect a system may, for example, be insufficiently motivated to do so, because the costs of failure fall on others. Such problems, Mr Anderson argues, are best examined using economic concepts, such as externalities, asymmetric information, adverse selection and moral hazard.

A classic example is that of fraud involving cash dispensers (automated teller machines). Mr Anderson investigated a number of cases of "phantom withdrawals", which customers said they never made, at British banks. He concluded that almost every time the security technology was working correctly, and that misconfiguration or mismanagement of the machines by the banks was to blame for the error. In Britain, it is customers, not banks, that are liable when phantom withdrawals are made, so the banks had little incentive to improve matters. In America, by contrast, it is the banks that are liable, so they have more of an incentive to train staff properly and install additional anti-fraud measures, such as cameras.

Similar examples abound on the internet. Suppose an attacker breaks into company A's computers and uses them to overload company B's computers with bogus traffic, thus keeping out legitimate users. Company B has suffered, in part, because of the insecurity of company A's systems. But short of a lawsuit from company B, company A has no incentive to fix the problem. Some examples of this sort of thing have already started to appear. In one case, a Texas judge issued a restraining order against three companies whose computers were being used by intruders to attack another firm's systems. The three companies were forced to disconnect from the internet until they could demonstrate that the vulnerabilities exploited by the attackers had been fixed.

Economic and legal measures will, predicts Mr Schneier, play an increasing role in compensating for perverse incentives that foster insecurity. Just as chief financial officers are legally required to sign statements declaring that company accounts are accurate, he speculates that, at least in certain industries, chief security officers might eventually have to sign security declarations. Similarly, product-liability lawsuits against software companies whose products are insecure would almost certainly discourage software-makers from cutting corners on security.

The enemy within

Incompetence and indifference are one thing; misconduct is another. Although external attacks get more attention in the media, notes a report from Vista Research, a consultancy, "the bulk of computer-security-related crime remains internal". Mr Anderson puts it a different way: the threat of hackers, he says, is "something that the security manager waves in your face to get the budget to deal with internal fraud". Vista estimates that 70% of security breaches that involve losses above $100,000 are perpetrated internally, often by disgruntled employees.

Attacks by insiders are potentially far costlier than external ones. The CSI/FBI survey, albeit using a small sample size, found that an insider attack against a large company caused an average of $2.7m-worth of damage, whereas the average external attack cost $57,000. A survey carried out by Oracle found that British companies believe malicious attacks by insiders pose more of a threat than external ones.

Defences against external attacks may not be much use against insiders. For a start, such people are likely to be inside the firewall (although companies are increasingly using internal firewalls between departments). And to an intrusion-detection system, an insider attack looks very different from an external one; by one estimate, an IDS has less than a 40% chance of distinguishing an insider attack from legitimate use of the network. One option is to use an analysis and visualisation tool, such as that made by SilentRunner. It represents network activity graphically to help security staff spot unusual behaviour – perhaps a large number of file transfers in a department where lay-offs have just been announced.

An alternative approach when fraud is suspected is to use "honey-pots" – decoy servers that lure attackers and collect evidence so that people who are up to no good can be identified. In one case cited by Recourse Technologies, a security firm that is now part of Symantec, a large financial firm discovered that its payroll systems had been compromised. Two dozen honeypots were set up, with names such as "payroll server", which caught the company's chief operating officer as he was trying to manipulate another executive's payroll record. He confessed to attempted fraud and resigned.

But the difficulty of combating insider attacks with technical means demonstrates that security is mainly a people problem. Indeed, the root cause of an insider attack may be poor management. An employee may resent being demoted or passed over for promotion, or feel underpaid or undervalued. Better management is a far more promising way to deal with these kinds of problems than technology.

The best way to prevent criminal activity by insiders is to make it difficult. "One of the key things you need is a separation of duties, so that no one individual runs everything," says Mr Spinks. Another simple measure is to ensure that all employees go on holiday at some point, to prevent them from maintaining tainted systems or procedures. Access privileges to company systems need to match employees' job descriptions so that, for example, only people in the personnel department can access employee records. When employees leave the company or their roles change, their access privileges must be revoked or altered immediately. And clear rules are needed to make sure that security staff know what to do if they detect abuse by senior managers. Better internal security procedures to deal with malicious insiders should also help to protect against external attacks, says Bill Murray of TruSecure.

One of the biggest threats to security, however, may be technological progress itself, as organisations embrace new technologies without taking the associated risks into account. To maintain and improve security, you need more than just the right blend of technology, policy and procedure. You also need to keep your eye on the ball as new technologies and new threats emerge.

Biometric fact and fiction

Body-scanning technology has its drawbacks

YOU'VE SEEN THEM IN SPY FILMS and science-fiction movies: eye-scanners, fingerprint readers, facial-recognition systems. Such body-scanning or "biometric" systems, which can make sure that somebody really is who he claims to be, are touted as the ultimate in security technology. Systems protected by passwords are unlocked by something you know (the password), which others can find out. Systems protected by keys or their high-tech equivalents, smart cards, are unlocked by something you have (the key), which others can steal. But systems protected by biometrics can be unlocked only by a bodily characteristic (such as a fingerprint) that no one can take from you. Your body is your password.

Eye-scanning biometric technology played a prominent part in a science-fiction movie, _Minority Report_. Its star, Tom Cruise, played a policeman accused of a crime who goes on the run. In the movie's futuristic setting, eye scanners are used to ensure that only legitimate users can access computer systems. Mr Cruise's character has eye transplants to conceal his identity, but also keeps his old eyeballs so that he can continue to log on to the police network.

That excursion into a fictional future highlights two real problems. The first is that the technology is not as secure as its proponents claim. Scanners that read fingerprints, the most widely used form of biometrics, proved easy to defeat in experiments carried out by Tsutomu Matsumoto, a security researcher at Yokohama National University. Mr Matsumoto was able to fool them around 80% of the time using fingers made of moulded gelatin. He was also able to take a photograph of a latent fingerprint (from a wine glass, for example) and use it to make a gelatin finger that fooled scanners 80% of the time as well. One advantage of gelatin is that having got past the guards, an intruder can eat the evidence.

Facial recognition, in which a computer analyses images from a digital camera and compares them with a "watch list" of known faces, is unreliable too. A study carried out at America's Defence Department found that instead of the claimed 90% accuracy rate, such systems correctly identified people only 51% of the time. Since the September 11th attacks, the technology has been tested at a number of American air-ports, but in one trial it was found that face-scanners could be fooled by people who turned their heads slightly. Recalibrating the system to allow looser matches caused a flood of false positives (where someone is wrongly identified as being on the watch list).

Identix, a leading supplier of facial-recognition systems, claims that its equipment's accuracy rate can be as high as 99%. But Mr Schneier, a security expert, says that even with an accuracy rate of 99.99%, and assuming that one in 10m fliers is a suspect whose face is on the watch list, there will still be 1,000 false alarms for every suspect identified. And most terrorists are not on watch lists. Face-scanning may reassure people and may have a deterrent effect, but these meagre benefits do not justify the costs.

The second and more important problem is that biometric technol-ogy, even when it works, strengthens only one link in the security chain. Its effectiveness is easily undermined by failures of process or policy. Tom Cruise's character in *Minority Report* is still able to get into the police computer network while on the run because someone has neglected to revoke his access privileges. This simple failure of process is all too common in real life. Another such real-world failure involves the use of hand-geometry scanners in airports. Each person's hand is supposed to be scanned separately, but often the first person in a group goes through the door and then holds it open.

In short, biometrics are no panacea. The additional security they pro-vide rarely justifies the cost. And in high-risk environments, such as banks or jails, other measures are still needed.

When the door is always open

The more that companies open up and interconnect their networks, the bigger the risk of security problems

A T THE HEIGHT OF THE DOTCOM BOOM, you could chart the rise and fall of companies by looking at the garish artwork sprayed on the walls of loft buildings in San Francisco's Multimedia Gulch district. But now, thanks to wireless technology, there is a better way. Driving around the city on a warm night, Bill Cockayne, a Silicon Valley veteran, opens his car's sunroof. His friend Nathan Schmidt posts what looks like a small fluorescent tube through the open roof and plugs it into a laptop computer. "Metro/Risk", says the computer in a clipped female voice as the car makes its way through North Beach. "Admin network. BCG." Then a robotic male voice booms out: "Microsoft WLAN. Archangel. Whistler. Rongi."

These are the names of computer networks in offices and homes that have been fitted with wireless access-points, which can provide internet access to users within range (typically, within 100 metres or so). Mr Schmidt's computer is configured so that open access-points, which can often be used by anyone within range, have their names spoken by a female voice; closed ones, for which a password is required, are read out by a male voice. Most of them are open. Mr Cockayne pulls over, and Mr Schmidt connects to a nearby access-point and calls up *The Economist*'s web page.

This kind of wireless networking, using the so-called Wi-Fi protocol, has become immensely popular (see Chart 2.5). Many companies and individuals leave their access-points open deliberately to enable passers-by to share their internet connections. Open a laptop in New York, San Francisco, Seattle or many other large cities around the world and you may well be able to get online free. But although Wi-Fi is liberating for users, it can cause security problems.

Adding an access-point to a network costs less than $200 and is very simple – so simple, in fact, that "rogue" access-points have started to sprout on corporate networks without the knowledge of management. A survey by *Computerworld*, an industry magazine, found that 30% of American companies had identified rogue access-points on their networks. And if these are left open, they provide a back door past the firewall into the company's network. Rob Clyde, chief technology officer at Symantec, says that half of the chief executives at a round-table event cited Wi-Fi as a top security concern.

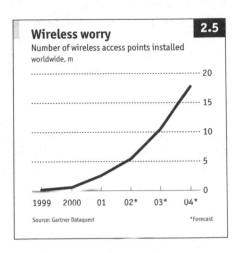

Wireless worry | 2.5
Number of wireless access points installed worldwide, m

1999 2000 01 02* 03* 04*

Source: Gartner Dataquest   *Forecast

This is just one example of how a new technology can bring security problems in its wake. There are plenty of others. Some firms are opening up their networks through online business-to-business exchanges, for example, where they list what they want to buy or sell and invite bids. Everything from paper clips to car components is bought or sold in this way. There is widespread agreement that "web services", in which companies open up their core business processes directly to other firms over the internet, will become increasingly important in the next few years. But by opening its systems to outsiders, a company may also attract unwanted visitors, or attacks from nosy competitors.

Joint ventures, in which two firms collaborate and share information, can also cause problems. A report by Vista Research cites the example of an American carmaker that established a joint venture with a Japanese firm and opened up its network to allow in employees of its Japanese partner. But the design of the American firm's network allowed access only on an "all or nothing" basis, so the Japanese firm's employees ended up with access to everything.

Handheld computers are another problem. They are often used to store sensitive data such as passwords, bank details and calendars. "The calendar is a fundamental loophole," says Doug Dedo of Microsoft's mobile devices division, because it may contain entries such as "meeting with company X re merger". Another problem associated with handheld computers is that their users carry them into the office and plug them into their computers, bypassing anti-virus systems and firewalls. A virus-infected document stored on a handheld computer could then start spreading. Similarly, peer-to-peer file-swapping networks such as Gnutella, instant-messaging services that zap messages and files across the internet, and web-based e-mail systems such as Hotmail all provide new routes into a company's network that can be exploited by attackers.

There are plenty of technical fixes available. Handheld scanners can be used to track down rogue access-points, and legitimate access-points can be secured against outsiders by using virtual-private-network (VPN) software. A lot of work is being done to ensure that web services are secure, including, improbably, a joint initiative by rivals Microsoft and IBM. Anti-virus and firewall software exists for handheld computers, which can also be password-protected. And firewalls can be configured to prevent unauthorised use of peer-to-peer and instant-messaging services.

All these threats arise from a common factor: the distinction between the "public" parts of a company's network (such as the web servers where its home page resides) and the private core (which is accessible only to employees) is quickly eroding. "The cultural and technological trend is towards more porous companies," says Gene Hodges, president of Network Associates, a large security-software firm. As firms connect with their suppliers and customers, "the more you open up, the more you are exposed".

Airports, not castles

The classic notion of perimeter security, in short, is fast becoming obsolete. Alan Henricks, chief executive of Cenzic, says the shift is "from keeping people out to bringing people in in a trusted fashion". Nand Mulchandani, co-founder of Oblix, another security firm, puts it more colourfully: the "big walls, moat and crocodiles" approach of the past few years, he says, is now outdated.

The latest thinking is that rather than seeing their networks as castles, large organisations should regard them as airports. People go in and out all the time, some areas are more secure than others, and as people pass from one area to another they have to present their credentials: tickets, boarding passes or passports. Apply this approach to computer security, and instead of an "exclusive" model in which you try to prevent people from doing things they shouldn't, you have an "inclusive" model that lays down who can do what, and only lets certain people do certain things.

In the old days, says Tony Scott, chief technology officer at General Motors, computer systems were used only internally, and managing who was allowed to do what was simple. But with the recent proliferation of systems, and a greater reliance on suppliers and outsourcing, the number of users who may need access to a company's systems has grown rapidly. "On top of that, most modern companies now have their actual business processes deeply embedded in their systems," he says. Indeed, their business processes are the systems. According to Mr Scott, "All these forces working together create a huge problem. Who is accessing these systems, and how can I manage it?"

One outfit offering solutions to this identity-management problem is Silicon-Valley-based Oblix. Its software sits between users and a company's existing software systems (accounts, inventory, e-mail, and so on). Using a big database that includes information on who can do what, it makes sure that users can do only the things they are meant to do.

It sounds obvious, but it has two advantages: it means users need to log in only once, rather than into lots of separate systems; and it centralises and simplifies the management of user privileges. For example, a division manager who hires or fires an employee can instantly update that employee's access privileges, rather than having to ask the systems department to make changes to a number of separate systems.

Responsibility for security can thus be devolved to managers and form part of their everyday management duties. Management is all-important, says Mr Mulchandani, because if your eyeball reader correctly identifies a sacked employee but his access privileges have not been revoked, you have a security failure on your hands. Oblix's software is used by a number of large firms including General Motors, Boeing and Pfizer. Identity-management systems are also available from other vendors, including Novell, IBM and ActivCard, whose smart-card-based offering is used by America's armed forces. The technique does not do away with the need for traditional security measures, but it provides an additional line of defence, particularly for large organisations that have to deal with a lot of users.

More importantly, identity management is an example of how technology can be used to align security procedures with business processes. Security thus becomes the servant of management. Security decisions must ultimately be taken by managers, not technical staff. The big decision, and the most difficult to make, is how much time and money to spend on security in the first place.

Putting it all together

Security spending is a matter of balancing risks and benefits

TOTAL COMPUTER SECURITY is impossible. No matter how much money you spend on fancy technology, how many training courses your staff attend or how many consultants you employ, you will still be vulnerable. Spending more, and spending wisely, can reduce your exposure, but it can never eliminate it altogether. So how much money and time does it make sense to spend on security? And what is the best way to spend them?

There are no simple answers. It is all a matter of striking an appropriate balance between cost and risk – and what is appropriate for one organisation might be wrong for another. Computer security, when you get down to it, is really about risk management. Before you can take any decisions about security spending, policy or management, the first thing you have to do is make a hard-headed risk assessment.

First, try to imagine all of the possible ways in which security could be breached. This is called "threat modelling", and is more difficult than it seems. Mr Schneier, a security guru, illustrates this point by asking people to imagine trying to eat at a pancake restaurant without paying. The obvious options are to grab the pancakes and run, or to pay with a fake credit card or counterfeit cash. But a would-be thief could devise more creative attacks.

He could, for example, invent some story to persuade another customer who had already paid for his meal to leave, and then eat his pancakes. He could impersonate a cook, a waiter, a manager, a celebrity or even the restaurant owner, all of whom might be entitled to free pancakes. He might forge a coupon for free pancakes. Or he might set off the fire alarm and grab some pancakes amid the ensuing chaos. Clearly, keeping an eye on the pancakes and securing the restaurant's payment system is not enough. Threat modelling alerts you to the whole range of possible attacks.

The next step is to determine how much to worry about each kind of attack. This involves estimating the expected loss associated with it, and the expected number of incidents per year. Multiply the two together, and the result is the "annual loss expectancy", which tells you how seriously to take the risk. Some incidents might cause massive losses, but be very rare; others will be more common, but involve smaller losses.

The final step is to work out the cost of defending against that attack. There are various ways to handle risk: mitigation (in the form of preventive technology and policies), outsourcing (passing the risk to someone else) and insurance (transferring the remaining risk to an insurer). Suppose you are concerned about the risk of your website being attacked. You can mitigate that risk by installing a firewall. You can outsource it by paying a web-hosting firm to maintain the website on your behalf, including looking after security for you. And you can buy an insurance policy that, in the event of an attack, will pay for the cost of cleaning things up and compensate you for the loss of revenue. There are costs associated with each of these courses of action. To determine whether a particular security measure is appropriate, you have to compare the expected loss from each attack with the cost of the defence against it.

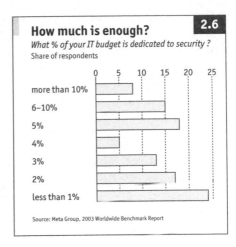

How much is enough? 2.6
What % of your IT budget is dedicated to security ?
Share of respondents

Source: Meta Group, 2003 Worldwide Benchmark Report

Firewalls make sense for large e-commerce websites, for example, because the cost of buying and maintaining a firewall is small compared with the revenue that would be lost if the site were shut down by an intruder, however briefly. But installing biometric eye-scanners at every turnstile on a city's public-transport system would be overkill, because fare-dodging can be mitigated with far cheaper technology. By contrast, in high-security environments such as military facilities or intelligence organisations, where a security breach would have serious consequences, the use of expensive security technology may be justified. In some situations, however, the right response may be to do nothing at all.

Standards stuff

That different organisations have different security needs is explicitly recognised in the ISO 17799, an international standard for "best practices in information security" that was introduced by the International Organisation for Standardisation in 2000. Risk analysis is a basic requirement of the standard, as is the establishment of a security policy. But, says Geoff Davies of i-Sec, a British security consultancy, "an industrial firm and a bank with ISO 17799 certification will have totally different systems." The standard does not specify particular technological or procedural approaches to security, but concentrates on broadly defined ends rather than specific means. The standard's flexibility is controversial, however. Critics believe future versions of the standard should be more prescriptive and more specific about what constitutes "best practice". Still, even in its current form, ISO 17799 is better than nothing. Many multinational companies have already embraced it to demonstrate their commitment to security. And in several Asian countries, companies that want to do business with the government electronically must conform to the standard.

Just as different organisations require different levels of protection, they will also respond to an attack in different ways. A large company, for example, may find it useful to have a dedicated security-response team. Scott Charney at Microsoft says that when an attack occurs, one of the things the team has to decide is whether to give priority to remediation or to investigation. Blocking the attack will alert the attacker, which may make collecting evidence against him difficult; but allowing the attacker to continue so that he can be identified may cause damage. Which is more appropriate depends on the context. In a military setting, tracking down the attacker is crucial; for a dotcom under attack by a teenager, blocking the attack makes more sense. Another difficult choice, says Mr Charney, is whether to bring in the police. Internal investigations allow an organisation to maintain control and keep things quiet, but law-enforcement agencies have broader powers.

For small and medium-sized companies, a sensible choice may be "managed security monitoring" (MSM). Firms that offer this service install "sentry" software and machines on clients' networks which relay a stream of messages to a central secure operations centre. Human operators watch for anomalous behaviour and raise the alarm if they detect anything suspicious. Using highly trained specialists to look out for trouble has the advantage that each operator can watch many networks at once, and can thus spot trends that would otherwise go unnoticed.

Risk analysis, and balancing cost and risk, is something the insurance industry has been doing for centuries. The industry is now showing increased interest in offering cover for computer-related risks. In the past, computer risks were included in general insurance policies, but were specifically excluded in the run-up to the year 2000 to avoid "millennium bug" liabilities. Now insurers are offering new products to protect companies against new risks. Because of the internet, "the landscape has changed," says David O'Neill, vice-president of e-Business Solutions at Zurich North America, which acts as a matchmaker between customers and underwriters. Greater connectivity means firms are now exposed to risks that were never contemplated by traditional insurance policies, he says.

Mr O'Neill can arrange insurance against a range of risks, including data theft, virus attacks or intrusions by malicious hackers, and loss of income owing to a security breach or network failure. Companies can also take out insurance against having to pay damages if confidential financial or medical data are accidentally or maliciously released. Because no two networks or businesses are alike, each policy is prepared individually.

Such cyber-insurance is, however, still very much in its infancy. The main problem is that the complexity of computer networks makes it very difficult to quantify risk accurately. By comparison, working out the likelihood that a 45-year-old smoker will have a heart attack in the next 12 months is a piece of cake. One reason for the lack of data, says Mr Charney, is that most security breaches are not detected or reported. But this will change. "When a company asks for $1m in damages after a virus outbreak, the insurer will say, 'Prove it'," he explains. "Firms will have to substantiate it, and we will get some data."

Mr Schneier predicts that insurance companies will start to specify what kinds of computer equipment companies should use, or charge lower premiums to insure more secure operating systems or hardware. Already, firms that use the monitoring service provided by his company, Counterpane Internet Security, enjoy a 20–40% reduction in their premiums for cyber-insurance. But Mr Anderson at Cambridge University thinks the need for cyber-insurance is overblown. "Insurers are having a hard time, so they are turning e-risks into a new pot of gold," he says.

Wrong department

Most organisations already have the expertise required to handle computer security in a sensible way. Usually, however, this risk-management expertise is found not in the systems department but in the finance department. "Chief information officers, chief financial officers and other executives already know how to do risk analysis," says Mr Davies. The systems department, however, does not; instead, it tends to be seduced by siren songs about technological fixes.

Enthusiasm for technological solutions can go too far. In two areas in particular, security technology could end up doing more harm than good. First, some measures introduced in the name of improving security may have the side-effect of needlessly infringing civil liberties. Face-scanning systems at airports are a good example. They are almost useless at spotting terrorists, but civil-rights advocates worry about "function creep", in which such systems are installed for one purpose and then used for another.

Similarly, new legislation is often proposed that would allow far more widespread wire-tapping and interception of internet communications to combat terrorism. But would it actually improve security? "Broad surveillance is generally the sign of a badly designed system of security," says Mr Schneier. He notes that the failure to predict the September 11th attacks was one of data sharing and interpretation, not data collection. Too much eavesdropping might actually exacerbate the problem, because there would be more data to sift. It would be better to step up intelligence gathering by humans.

The second area where security technology could do more harm than good is in the world of business. Technology introduced to improve security often seems to have the side-effect of reinforcing the market dominance of the firm pushing it. "Information-security technologies are more and more used in struggles between one company and another," says Mr Anderson. "Vendors will build in things that they claim are security mechanisms but are actually there for anti-competitive reasons."

One highly controversial example is Palladium, Microsoft's proposed technology for fencing off secure areas inside a computer. It might be very useful for stopping viruses; but it might also enable Microsoft to gain control of the standard for the delivery of digital music and movies. (See Postscript on page 76.)

Security, in sum, depends on balancing cost and risk through the appropriate use of both technology and policy. The tricky part is defining what "appropriate" means in a particular context. It will always be a balancing act. Too little can be dangerous and costly – but so can too much.

The mouse that might roar

Cyber-terrorism is possible, but not very likely

IT IS A DEVASTATING PROSPECT. Terrorists electronically break into the computers that control the water supply of a large American city, open and close valves to contaminate the water with untreated sewage or toxic chemicals, and then release it in a devastating flood. As the emergency services struggle to respond, the terrorists strike again, shutting down the telephone network and electrical power grid with just a few mouse clicks. Businesses are paralysed, hospitals are overwhelmed and roads are gridlocked as people try to flee.

This kind of scenario is invoked by doom-mongers who insist that stepping up physical security since the September 11th attacks is not enough. Road-blocks and soldiers around power stations cannot prevent digital terrorism. "Until we secure our cyber-infrastructure, a few keystrokes and an internet connection are all one needs to disable the economy and endanger lives," Lamar Smith, a Texas congressman, told a judiciary committee in February 2002. He ended with his catchphrase: "A mouse can be just as dangerous as a bullet or a bomb." Is he right?

It is true that utility companies and other operators of critical infrastructure are increasingly connected to the internet. But just because an electricity company's customers can pay their bills online, it does not necessarily follow that the company's critical control systems are vulnerable to attack. Control systems are usually kept entirely separate from other systems, for good reason. They tend to be obscure, old-fashioned systems that are incompatible with internet technology anyhow. Even authorised users require specialist knowledge to operate them. And telecoms firms, hospitals and businesses usually have contingency plans to deal with power failures or flooding.

A simulation carried out in August 2002 by the United States Naval War College in conjunction with Gartner, a consultancy, concluded that an "electronic Pearl Harbour" attack on America's critical infrastructure could indeed cause serious disruption, but would first need five years of preparation and $200m of funding. There are far simpler and less costly ways to attack critical infrastructure, from hoax phone calls to truck bombs and hijacked airliners.

In September 2002 Richard Clarke, America's cyber-security tsar, unveiled his long-awaited blueprint for securing critical infrastructure from digital attacks. It was a bit of a damp squib, making no firm recommendations and proposing no new regulation or legislation. But its lily-livered approach might, in fact, be the right one. When a risk has been overstated, inaction may be the best policy.

It is difficult to avoid comparisons with the "millennium bug" and the predictions of widespread computer chaos arising from the change of date to the year 2000. Then, as now, the alarm was sounded by technology vendors and consultants, who stood to gain from scaremongering. But Ross Anderson, a computer scientist at Cambridge University, prefers to draw an analogy with the environmental lobby. Like eco-warriors, he observes, those in the security industry – be they vendors trying to boost sales, academics chasing grants, or politicians looking for bigger budgets – have a built-in incentive to overstate the risks.

POSTSCRIPT

Since this section was published in 2002, digital security has remained a high priority for both companies and consumers. As technological defences have been strengthened against spam and viruses, scam artists have increasingly resorted to "social engineering" attacks such as "phishing", in which internet users are tricked by bogus e-mails into revealing financial information that can be used to defraud them. Microsoft now releases security patches to its software once a month, which makes it easier for systems administrators to keep systems up to date. But it has watered down plans for its "Palladium" technology, now known as Next Generation Secure Computing Base, as a result of technical problems and objections from software-makers and users.

The material on pages 42–76 first appeared in a survey in *The Economist* in October 2002.

3
MAKE IT SIMPLE

Make it simple

The next thing in technology is not just big but truly huge: the conquest of complexity

"THE COMPUTER KNOWS ME AS ITS ENEMY," says John Maeda. "Everything I touch doesn't work." Take those "plug-and-play" devices, such as printers and digital cameras, that any personal computer (PC) allegedly recognises automatically as soon as they are plugged into an orifice called a USB port at the back of the PC. Whenever Mr Maeda plugs something in, he says, his PC sends a long and incomprehensible error message from Windows, Microsoft's ubiquitous operating system. But he knows from bitter experience that the gist of it is no.

At first glance, Mr Maeda's troubles might not seem very noteworthy. Who has not watched Windows crash and reboot without provocation, downloaded endless anti-virus programs to reclaim a moribund hard disc, fiddled with cables and settings to hook up a printer, and sometimes simply given up? Yet Mr Maeda is not just any old technophobic user. He has a master's degree in computer science and a PhD in interface design, and is currently a professor in computer design at the Massachusetts Institute of Technology (MIT). He is, in short, one of the world's foremost computer geeks. Mr Maeda concluded that if he, of all people, cannot master the technology needed to use computers effectively, it is time to declare a crisis. So in 2004 he launched a new research initiative called "Simplicity" at the MIT Media Lab. Its mission is to look for ways out of today's mess.

Mr Maeda has plenty of sympathisers. "It is time for us to rise up with a profound demand," declared the late Michael Dertouzos in his 2001 book, *The Unfinished Revolution* (HarperBusiness): "Make our computers simpler to use!" Donald Norman, a long-standing advocate of design simplicity, concurs. "Today's technology is intrusive and overbearing. It leaves us with no moments of silence, with less time to ourselves, with a sense of diminished control over our lives," he writes in his book, *The Invisible Computer* (MIT Press, 1998). "People are analogue, not digital; biological, not mechanical. It is time for human-centred technology, a humane technology."

The information-technology (IT) industry itself is long past denial. Greg Papadopoulos, chief technologist at Sun Microsystems, a maker of powerful corporate computers, says that IT today is "in a state that we should be ashamed of; it's embarrassing". Ray Lane, a venture capitalist at Kleiner Perkins Caufield & Byers, one of the most prominent technology financiers in Silicon Valley, explains: "Complexity is holding our industry back right now. A lot of what is bought and paid for doesn't get implemented because of complexity. Maybe this is the industry's biggest challenge." Even Microsoft, which people like Mr Lane identify as a prime culprit, is apologetic. "So far, most people would say that technology has made life more complex," concedes Chris Capossela, the boss of Microsoft's desktop applications.

The economic costs of IT complexity are hard to quantify but probably exorbitant. The Standish Group, a research outfit that tracks corporate IT purchases, has found that 66% of all IT projects either fail outright or take much longer to install than expected because of their complexity. Among very big IT projects – those costing over $10m apiece – 98% fall short.

Gartner, another research firm, uses other proxies for complexity. An average firm's computer networks are down for an unplanned 175 hours a year, calculates Gartner, causing an average loss of over $7m. On top of that, employees waste an average of one week a year struggling with their recalcitrant PCs. And itinerant employees, such as salesmen, incur an extra $4,400 a year in IT costs, says the firm.

Tony Picardi, a boffin at IDC, yet another big research firm, comes up with perhaps the most frightening number. When he polled a sample of firms at the beginning of the 1990s, they were spending 75% of their IT budget on new hardware and software and 25% on fixing the systems that they already had; now that ratio has been reversed – 70–80% of IT spending goes on fixing things rather than buying new systems. According to Mr Picardi, this suggests that in 2004 alone IT complexity cost firms worldwide some $750 billion. Even this, however, does not account for the burden on consumers, whether measured in the cost of call-centres and help desks, in the amount of gadgets and features never used because they are so byzantine, or in sheer frustration.

Why now?

Complaints about complex technology are, of course, nothing new. Arguably, IT has become more complex in each of the 45 years since the integrated circuit made its debut. But a few things have happened in the past few years that now add a greater sense of urgency.

The most obvious change is the IT bust that followed the dotcom

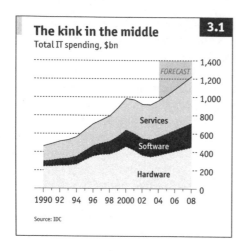

The kink in the middle `3.1`
Total IT spending, $bn

FORECAST

Services

Software

Hardware

1,400
1,200
1,000
800
600
400
200
0

1990 92 94 96 98 2000 02 04 06 08

Source: IDC

boom of the late 1990s. After a decade of strong growth, the IT industry suddenly started shrinking in 2001 (see Chart 3.1). In early 2000 it accounted for 35% of America's S&P 500 index; by 2004 its share was down to about 15%. "For the past few years, the tech industry's old formula – build it and they come – has no longer worked," says Pip Coburn, a technology analyst at UBS, an investment bank. For technology vendors, he thinks, this is the sort of trauma that precedes a paradigm shift. Customers no longer demand "hot" technologies, but instead want "cold" technologies, such as integration software, that help them stitch together and simplify the fancy systems they bought during the boom years.

Steven Milunovich, an analyst at Merrill Lynch, another bank, offers a further reason why simplicity is only now becoming a big issue. He argues that the IT industry progresses in 15-year waves. In the first wave, during the 1970s and early 1980s, companies installed big mainframe computers; in the second wave, they put in PCs that were hooked up to "server" computers in the basement; and in the third wave, which is breaking now, they are beginning to connect every gadget that employees might use, from hand-held computers to mobile phones, to the internet.

The mainframe era, says Mr Milunovich, was dominated by proprietary technology (above all, IBM's), used mostly to automate the back offices of companies, so the number of people actually working with it was small. In the PC era, de facto standards (ie, Microsoft's) ruled, and technology was used for word processors and spreadsheets to make companies' front offices more productive, so the number of people using technology multiplied tenfold. And in the internet era, Mr Milunovich says, de jure standards (those agreed on by industry consortia) are taking over, and every single employee will be expected to use technology, resulting in another tenfold increase in numbers.

Moreover, the boundaries between office, car and home will become increasingly blurred and will eventually disappear altogether. In rich countries, virtually the entire population will be expected to be permanently connected to the internet, both as employees and as consumers. This will at last make IT pervasive and ubiquitous, like electricity or telephones before it, so the emphasis will shift towards making gadgets and networks simple to use.

UBS's Mr Coburn adds a demographic observation. Today, he says, some 70% of the world's population are "analogues", who are "terrified by technology", and for whom the pain of technology "is not just the time it takes to figure out new gadgets but the pain of feeling stupid at each moment along the way". Another 15% are "digital immigrants", typically thirty-somethings who adopted technology as young adults; and the other 15% are "digital natives", teenagers and young adults who have never known and cannot imagine life without IM (instant messaging, in case you are an analogue). But a decade from now, Mr Coburn says, virtually the entire population will be digital natives or immigrants, as the ageing analogues convert to avoid social isolation. Once again, the needs of these converts point to a hugely increased demand for simplicity.

The question is whether this sort of technology can ever become simple, and if so, how. This section analyses the causes of technological complexity both for firms and for consumers, evaluates the main efforts toward simplification by IT and telecom vendors today, and considers what the growing demands for simplicity mean for these industries. A good place to start is in the past.

Now you see it, now you don't

To be truly successful, a complex technology needs to "disappear"

THERE HAS NEVER BEEN ANYTHING quite like information technology before, but there have certainly been other complex technologies that needed simplifying. Joe Corn, a history professor at Stanford University, believes that the first example of a complex consumer technology was clocks, which arrived in the 1820s. Clocks were sold with user manuals, which featured entries such as "How to erect and regulate your device". When sewing machines appeared in the 1840s, they came with 40-page manuals full of detailed instructions. Discouragingly, it took two generations until a trade publication was able to declare in the 1880s that "every woman now knows how to use one."

At about the same time, the increase in technological complexity gathered pace. With electricity came new appliances, such as the phonograph, invented in 1877 by Thomas Alva Edison. According to Donald Norman, a computer-design guru, despite Mr Edison's genius for engineering he was a marketing moron, and his first phonograph was all but unusable (in fact, initially he had no particular uses in mind for it). For decades, Mr Edison fiddled with his technology, always going for the most impressive engineering solution. For instance, he chose cylinders over discs as the recording medium. It took a generation and the entry of a new rival, Emile Berliner, to prepare the phonograph for the mass market by making it easier to use (introducing discs instead of cylinders) and giving it a purpose (playing music). Mr Edison's companies foundered whereas Mr Berliner's thrived, and phonographs became ubiquitous, first as "gramophones" or "Victrolas", the name of Mr Berliner's model, and ultimately as "record players".

Another complex technology, with an even bigger impact, was the car. The first cars, in the early 1900s, were "mostly a burden and a challenge", says Mr Corn. Driving one required skill in lubricating various moving parts, sending oil manually to the transmission, adjusting the spark plug, setting the choke, opening the throttle, wielding the crank and knowing what to do when the car broke down, which it invariably did. People at the time hired chauffeurs, says Mr Corn, mostly because they needed to have a mechanic at hand to fix the car, just as firms today need IT staff and households need teenagers to sort out their computers.

By the 1930s, however, the car had become more user-friendly and ready for the mass market. Two things in particular had made this possible. The first was the rise, spread and eventual ubiquity of a support infrastructure for cars. This included a network of decent roads and motorways, and of petrol stations and garages for repair. The second was the makers' increasing skill at hiding the technology from drivers. Ford proved particularly good at this. Ironically, it meant that cars got hugely more complex on the inside, because most of the tasks that had previously been carried out by drivers now had to be done automatically. This presented drivers with a radically simplified surface, or "interface" in today's jargon, so that all they had to do was turn the ignition key, put their foot on the accelerator, brake, steer and change gear – and after 1940, when automatic transmissions were introduced, even gear-shifting became optional.

Another instructive technology is electricity. In its early days, those firms and households that could afford it had their own generators. Keeping these going soon became a full-time job. In the early 20th century, writes Nick Carr, the author of a book entitled *Does IT Matter?* (Harvard Business School Press, 2004), most companies had a senior management position called "vice-president of electricity", a rough equivalent of today's "chief information officer" (CIO) and "chief technology officer" (CTO). Within a generation, however, the generators and vice-presidents disappeared as electricity became available through the grid, leaving users to deal only with the simplest of interfaces, the power socket.

Out with the nerds

The evolution of these technologies holds some lessons for the IT industry today. The first observation, according to Mr Norman, "is that in the early days of any technological revolution the engineers are in charge, and their customers are the early adopters. But the mass market is the late adopters. This is why Thomas Alva Edison, an engineering genius, failed miserably in business." Similarly, in IT today, says Mr Papadopoulos of Sun Microsystems, "the biggest problem is that most of the people who create these artefacts are nerds. I want to see more artists create these things."

The geekiness that predominates in the early stages of any new technology leads to a nasty affliction that Paul Saffo, a technology visionary at California's Institute for the Future, calls "featuritis". For example, Microsoft in a recent survey found that most consumers use only 10% of the features on offer in Microsoft Word. In other words, some 90% of this software is clutter that obscures the few features people actually want. This violates a crucial principle of design. As Soetsu Yanagi wrote in *The Unknown Craftsman*, his classic 1972 book on folk art, "man is most free when his tools are proportionate to his needs". The most immediate problem with IT today, as with other technologies at comparable stages, says Mr Saffo, is that "our gadgets are so disproportionate".

A second lesson from history, however, is that a brute cull of features would be futile. As technologies, the sewing machine, the phonograph, the car and the electricity grid have only ever grown more complex over time. Today's cars, in fact, are mobile computers, containing dozens of microchips and sensors and other electronic sub-systems that Henry Ford would not recognise. Electricity grids today are as complex as they are invisible in everyday life. Consumers notice them only when things go wrong, as they did spectacularly during 2003's power cuts in northeastern America and Canada.

"You have to push all the complexity to the back end in order to make the front end very simple," says Marc Benioff, the boss of Salesforce.com, a software firm (see page 91). This migration of complexity, says Mr Benioff, echoes the process of civilisation. Thus, every house initially has its own well and later its own generator. Civilisation turns houses into "nodes" on a public network that householders draw on. But the "interface" – the water tap, the toilet flush, the power switch – has to be "incredibly simple". All the management of complexity now takes place within the network, so that consumers no longer even know when their electricity or water company upgrades its technology. Thus, from the user's point of view, says Mr Benioff, "technology goes through a gradual disappearance process".

From the point of view of the vendors, the opposite is true. "Our experience is that for every mouse click we take out of the user experience, 20 things have to happen in our software behind the scenes," says Brad Treat, the chief executive of SightSpeed, a company that wants to make video phone calls as easy for consumers as e-mailing. The same applies to corporate datacentres. "So don't expect some catharsis in eliminating layers of software," says Mr Papadopoulos. "The way we get rid of complexity is by creating new layers of abstraction and sedimenting what is below." This will take different forms for firms and for consumers. First, consider the firms.

A byte's-eye view of complexity

Companies' computer infrastructures contain a Pandora's boxful of trouble

ONE WAY TO APPRECIATE the chaotic complexity that rules in the computer vaults ("datacentres") of firms is to imagine, with a bit of anthropomorphic licence, the journey of one lowly unit of digital information, or byte, as it wends its way on a routine mission through a maze of computers, routers, switches and wires.

At the outset, the byte is asleep on a specialised storage disc. This disc could be made by a firm such as EMC or Hitachi. Now an alarm bell rings and a message flashes that an employee of the company, sitting in an office somewhere half-way round the world, has clicked on some button in his PC's software. The byte wakes up and is ejected from its storage disc. Along with billions of other bytes from other storage discs, it is now herded through a tunnel called a storage switch. This switch is probably made by a company called Brocade or McData. It hurls the byte towards an "interface card", which comes from yet another vendor, and the card directs the byte into one of the datacentre's many back-office computers, called "servers".

This causes the byte some momentary confusion, because the datacentre has servers that were assembled by different makers, such as IBM, Hewlett-Packard, Sun Microsystems or Dell. It also has some mainframes that are left over from an earlier era. Some of the servers will contain a microprocessor made by Intel, whereas others run on chips from AMD or on Sun's Sparc chip, and the mainframes are using IBM chips. For their operating system, some of the servers use Windows, others Linux or Solaris or a more obscure kind of Unix software, and the mainframes run on their own, proprietary, system.

The byte is catapulted into this motley and, with luck, finds the appropriate server. As it arrives in that machine, the byte is spun around by a layer of "virtualisation" software, which might come from a company called Veritas. This program gives the byte a quick health-check to see whether a copy needs to be retrieved from a back-up tape on another network, probably a long way away, set up to guard against disasters such as earthquakes. That tape probably comes from StorageTek.

When this is done, the server shoots the byte to another computer and into a database program. This database probably comes from Oracle or IBM. The byte then ricochets into yet another server computer that runs a whole bag of programs called "middleware", which might be written by BEA Systems or IBM or TIBCO. The middleware now hands the byte over to the application software of the employee who started this journey with his click. That application program could come from SAP, PeopleSoft, Oracle, Siebel or a number of other companies. Just as the byte arrives, dizzy and dazed, the employee clicks again, and another journey through the labyrinth begins.

Twisted tongues

But the poor byte not only has to navigate a labyrinth; it also has to cope with Babel. Every time it moves, it has to get past yet another sentry, called an interface, hired by whichever vendor was subcontracted to build that particular intersection. These sentries demand different passwords, called "protocols", and speak different languages. The byte, in other words, has to travel with a suitcase of dictionaries. With luck, it can make some progress by speaking a lingua franca such as Java or .NET, and by brandishing widely used passwords that are the internet equivalent of its mother's maiden name.

Sooner or later, however, the byte encounters some truly anachronistic sentries, called "legacies". According to estimates by *InfoWorld*, a trade publication, about half of all corporate data today still reside on mainframes, possibly bought decades ago. And many companies still use bespoke software that was written in the 1980s – before off-the-shelf, packaged software arrived – by the company's own IT staff, who left the company long ago and took their little secrets with them. The byte, in other words, also has to be fluent in Latin, Aramaic, Hittite and other extinct tongues to keep moving.

Along the way, moreover, it encounters open paranoia. Whenever it mingles with bytes that started their journey in the computers of another datacentre, it has to pass through checkpoints, called firewalls, that check its identification documents against a list of viruses and worms and other nasty bytes, roughly as the Israeli army might examine a Palestinian entering from the Gaza strip. In fact, the market leader for such firewalls is an Israeli firm called Check Point Software Technologies.

Occasionally, the byte will also get wrapped in several sealed layers of encryption and sent to its destination as a VPN, or "virtual private network", only to have to be carefully unwrapped again at the other end.

Throughout the journey, the byte will be shadowed by a digital Department of Homeland Security, called an "intrusion detection system" (IDS).

Over the years, every firm acquires an agglomeration of boxes and code as unique as a fingerprint. Then firms merge, and someone has to try to stitch several of these unique datacentres together. This is the sort of thing that Charles de Felipe did at J.P. Morgan, a huge global bank, where he was one of the chief technical people for 26 years until he quit in July 2004. During his career there Mr de Felipe went through nine mergers, which amalgamated once-famous names such as Chemical, Horizon, Manufacturers Hanover, Chase, H&Q, Jardine Fleming, J.P. Morgan and, most recently, BankOne into a single bank. "Every four years or so the entire landscape changes," says Mr de Felipe. "On day one you merge the books; on day two you do the regulatory paperwork, and on day three you start talking about the systems." The first two, he says, are child's play compared with the third. In his last few years on the job, for instance, he was concentrating mostly on reducing the number of the bank's desktop applications, from a total of 415 to about 40.

All this opens a Pandora's box of problems. Something in the datacentre will go wrong almost all the time. When that happens, the users will scream for the IT staff, who will have to figure out where in this chain of almost infinite permutations the byte got stuck or lost. There is software that can run a few tests. All too often, however, it comes back with the dreaded NTF ("no trouble found") message, says Kenny Gross, a researcher at Sun who came from the nuclear industry, "where meltdown is not a metaphor". That means the IT staff are reduced to changing devices one by one to find the villain. This can take days, weeks or months.

Today's datacentres are a catastrophic mess, says Alfred Chuang, the boss of BEA Systems, a middleware company that he co-founded a decade ago (he is the A in BEA), with the explicit aim of simplifying datacentres. The struggle between complexity and simplicity, he reckons, "is binary: Either it will all blow up, or it will simplify". For the moment, no one can tell. But remember that the last spirit left in Pandora's box, once all the evil ones had escaped, was Hope.

If in doubt, farm it out

The ultimate solution to simplifying your datacentre is not to have one at all

EVERY SELF-RESPECTING TECHNOLOGY VENDOR these days not only vigorously deplores complexity but also claims to have a solution, and a suitably dramatic name for it to boot. Thus, Hewlett-Packard (HP) talks about its vision for the "adaptive enterprise", helped by HP simplification software called OpenView. IBM trumpets the dawn of "on-demand" IT for companies through IBM's "autonomic computing" architecture. EDS, an IT consultancy, offers the "agile enterprise". Hitachi has "harmonious computing". Forrester, a research firm, suggests "organic IT". Sun tempts with a shrewdly mysterious name, "N1". Dell has "dynamic computing" and Microsoft flaunts the grand-sounding "dynamic systems initiative".

All these marketing buzzwords imply a promise to hide the complexity of firms' datacentres in the same way that modern cars and planes hide their technological complexity from drivers and pilots. This is hard to argue with. At the same time, the grand titles raise expectations to an exalted level. Words such as "organic" and "autonomic" intentionally invite comparisons with biological systems whose complexity is hidden from the creatures living within them. The implication is that digital technology can achieve the same feat.

Take, for instance, IBM's autonomic-computing initiative, launched in 2002 by Alan Ganek, an IBM executive, and now the most ambitious proposal on offer. The label is currently attached to about 50 distinct IBM products with over 400 product features. In the longer term, however, IBM is hoping to bring computing to a level where it mimics the autonomic nervous system of the human body. This is what regulates breathing, digestion, blood-sugar levels, temperature, pancreatic function, immune responses to germs, and so on automatically and without the people concerned being conscious of these processes. It is, in a way, nature's gold standard of virtualisation software and complexity concealment, which is why IBM bagged the metaphor.

What IBM actually means by "autonomic" in a computing context, Mr Ganek explains, comes down to four technological goals. The first is to make computers and networks "self-configuring". Whereas today IT staff walk around and manually perform tasks such as plugging CDs into computers or fiddling with command lines, IBM wants the hardware and software itself to figure out what settings are missing and to install them automatically.

The second step is to make the systems "self-healing". Thus, the network should diagnose problems automatically – for example, by noticing a crashed computer and rebooting it. Whereas today IT staff can easily take several weeks to diagnose a problem by manually sorting through logs, autonomic computing can get it done without human intervention in about 40 minutes, says Mr Ganek.

The third goal, Mr Ganek continues, is to make systems "self-optimising". This means that the network should know how to balance processing workloads among the various servers and storage computers so that none is idle or swamped. And the final step is to make the whole network "self-protecting". The system, in other words, should be able to anticipate, hunt down and kill computer viruses and worms all by itself; to tell spam from legitimate e-mail; and to prevent "phishing" and other data theft.

A pinch of salt

The vision is shockingly ambitious. If it ever becomes reality, IBM (or HP, or whoever gets there first) will in essence have achieved what it has taken millions of years of natural evolution to do in the analogue, biological world. Not surprisingly, many experts are sceptical, pointing to the parallel with artificial intelligence (AI), which boffins confidently described as imminent in the 1950s but which remains elusive to this day. Mr Coburn at UBS says the talk of autonomic computing reminds him "of a high-school science fair", and thinks it may be just another one of those things that IT vendors "throw on the wall to see what sticks".

Buried deep underneath the guff, however, there is indeed a technology widely considered to have the potential for radical simplification. Like the wheel, the zip fastener and other breakthrough technologies, it looks deceptively basic at first sight. Even its name, "web services", is so vague that vendors find it hard to build any hype for a lay audience around it.

The best way to understand web services is to stop thinking of either "webs" or "services" and instead to picture Lego blocks. These little Danish plastic toy bricks come in different colours, shapes and sizes, but all Lego blocks have the same standardised studs and corresponding holes that allow them to be assembled, taken apart and reassembled in all sorts of creative ways. The magic of web services, in effect, is to turn almost any fiddly piece in any chaotic datacentre into a Lego block, so that it can snugly fit together with all the other fiddly bits. Thus, data-centres that consist of decades of legacy systems and lots of incompatible machines can now be snapped together and apart, Lego by Lego.

In place of studs and holes, web services use standardised software that wraps itself around existing computer systems. These wrappers do several things. First, they describe what the component inside is and what it does. Then they post this description to a directory that other computers can browse. This allows those other computers – which can belong either to the same company or to independent suppliers and customers – to find and use the software inside the wrapper.

This removes the main bottleneck that scuppered business-to-business computing during the dotcom bubble. "The whole B2B boom died for one simple reason: nobody could get their damn systems to talk together," says Halsey Minor, the founder of Grand Central Communications, a start-up that uses web services to stitch datacentres together. Now, he says, they do talk together.

Imagine, for example, that a company receives an electronic order. The software application that takes these orders must first ensure that the customer has an adequate credit history. It therefore consults a directory of web services, finds an application from an independent firm that checks credit ratings, contacts this application and finds out that the customer is a reliable debtor. Next, the software consults the directory again, this time to find an internal application that keeps track of inventory in the warehouse, and finds that the product is in store. Now it goes back to the directory and looks for an external billing service, and so forth until the entire transaction is closed.

Making a splat

As a way of simplifying computing, web services have been talked about for some time. Only in the past couple of years, however, has there been real progress in agreeing on the most vital aspect, the standards that will make every system look familiar to everybody else. A major breakthrough came in October 2003, when the industry's two superpowers, Microsoft and IBM, got up on a stage together and stated what protocols they intend to use. Collectively dubbed "ws splat" in geeky circles, these are now being adopted by the rest of the industry.

This has raised hopes for a huge increase in their use in the next few years (see Chart 3.2). Ronald Schmelzer and Jason Bloomberg at ZapThink, a consultancy, think that web services are "nearing their tipping point", because they benefit from "the network effect: the adoption rate of the network increases in proportion to its utility". In other words, as with telephones or e-mail, a network with only a few people on it is not very useful; but as more people join it, it becomes exponentially more useful and thereby attracts even more members, and so on.

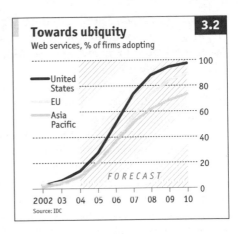

Towards ubiquity `3.2`

Web services, % of firms adopting

— United States
··· EU
— Asia Pacific

FORECAST

2002 03 04 05 06 07 08 09 10

Source: IDC

Taking the idea of web services to its logical extreme, it is reasonable to ask why firms should continue to amass their own piles of Lego blocks, most of which will only duplicate the Lego blocks of business partners. Put differently, why have a datacentre if all you want is the data? This is a fairly new idea in the IT industry, although in many established industries it has been around for a long time. People do not put safes into their basements but open bank accounts. Similarly, "most people shouldn't build their own aeroplanes," says Sun's Mr Papadopoulos. "They shouldn't even own them; in fact, they shouldn't even rent them; what they should do is rent a seat on one."

In IT, the equivalent of renting a seat on an aircraft is to rent software as a service from specialised firms called "application service providers", or ASPs. These companies build huge datacentres so that other companies do not have to. The best-known ASP today is Salesforce.com, a San Francisco firm that made its debut on the stockmarket in June 2004. As the name suggests, Salesforce.com specialises in software that salespeople use to keep track of their marketing leads and client information. Traditionally, firms buy this kind of software from vendors such as Siebel Systems, then try to integrate it into their own datacentres. With Salesforce.com, however, firms simply pay a monthly fee, from $65 per user, and go to Salesforce.com's website, just as they go to Amazon's when they want to shop for books, or eBay's to buy secondhand goods.

This arrangement makes a lot of things simpler. Users need to spend less time on training courses, because the interface – in essence, the web browser – is already familiar to them. "I can train the average customer in under 45 minutes on the phone," claims Marc Benioff, Salesforce.com's boss, adding that traditional software packages often take weeks to learn.

The IT staff of the firm using Salesforce.com also have less work to do. They do not have to install any new software on the firm's own computers, and can leave Salesforce.com to worry about integrating its software with the client's other systems. Even upgrading the software becomes much easier. Instead of shipping boxes of CDs to its customers, Salesforce.com simply shuts down its system for a few hours on a weekend night, and when clients log on again on Monday morning they see the new version in their browsers.

As an industry, ASPs got off to a bad start. The first generation, which sprang up during the dotcom boom, had trouble integrating their applications with their clients' legacy systems, and ended up recreating the complexity of their clients' datacentres in their own basements. When the dotcom bubble burst, says Mr Lane at Kleiner Perkins Caufield & Byers in Silicon Valley, those early ASPs collapsed "because we VCs wouldn't invest in them any more".

The second generation, however, seems to have cracked the problem of integration, thanks to web services, and is now picking off segments of the software market one by one. IDC estimates that ASPs' overall revenues will grow from $3 billion in 2003 to $9 billion by 2008. As Grand Central's Mr Minor sees it, that puts IT today on the same path as other technologies in history, as "complexity gets concentrated in the middle of the network, while the edge gets simple".

Spare me the details

There is a huge gap between what consumers want and what vendors would like to sell them

L ISA HOOK, an executive at AOL, one of the biggest providers of traditional ("dial-up") internet access, has learned amazing things by listening in on the calls to AOL's help desk. Usually, the problem is that users cannot get online. The help desk's first question is: "Do you have a computer?" Surprisingly often the answer is no, and the customer was trying to shove the installation CD into the stereo or TV set. The help desk's next question is: "Do you have a second telephone line?" Again, surprisingly often the answer is no, which means that the customer cannot get on to the internet because he is on the line to the help desk. And so it goes on.

Admittedly, in America, where about half of all internet households now have high-speed ("broadband") connections, these AOL customers are so-called "late adopters", or "analogues". But even younger, savvier "digital natives" or "digital immigrants" can provide surprising insights for those who care to listen.

Genevieve Bell, an anthropologist who works for Intel, the world's biggest semiconductor-maker, has been travelling around Asia for three years to observe how Asians use, or choose not to use, technology. She was especially struck by the differences in how westerners and Asians view their homes. Americans tended to say things like "my home is my castle" and furnish it as a self-contained playground, says Ms Bell. Asians were more likely to tell her that "my home is a place of harmony", "grace", "simplicity" or "humility". These Asians recoiled from gadgets that made noises or looked showy or intrusive.

Even within western cultures, Ms Bell, who is Australian, has found startling differences in the way people view technology. When she recently opened her laptop in a café in Sydney to check her e-mail on the local wireless network, using a fast-spreading technology called Wi-Fi, she immediately got a mocking "Oi, what do you think you are, famous?" from the next table. "For Americans, adopting technology is an expression of American-ness, part of the story of modernity and progress," says Ms Bell. For many other people, it may be just a hassle, or downright pretentious.

And even Americans, perhaps more prone than others to workaholism, can get frustrated by technology. Chris Capossela, boss of productivity software at Microsoft, commissioned a study where office workers were shadowed (with their consent) after they left the office. It showed that people feel pressure even in their cars and homes to keep up with "the expectation that one is always available," says Mr Capossela. Thanks to technology (laptops, BlackBerries, smart phones and so on), he says, "the boundaries of nine-to-five no longer exist". This creates a new demographic category, "the enterprise consumer", for whom not only technology but all of life has grown more complex.

Hark, the vendors

Contrast these insights with the technological visions that the industry is currently peddling. The best place to see them is the Consumer Electronics Show (CES), held every January in Las Vegas. For the better part of a week, some 133,000 visitors roam a space the size of several football fields and duck in and out of 2,500 exhibitors' booths. Almost all the visitors are male, and the toilets are blanketed with call girls' business cards. Everything else is a flashing warren of flat-panel screens, robots that serve drinks and countless other outlandish gadgets. The CES is where everybody who is anybody in consumer electronics, computing and telecoms comes to announce their new products.

A small portion of these wares eventually do end up being used by ordinary humans. Currently, the CES technophiles are excited about two trends in particular. The first is that every single electronic device will soon be connected to the internet. This includes the obvious, such as mobile phones and TV sets, and the less obvious, such as shirts and nappies that carry tiny radio-frequency identification (RFID) tags. Microsoft talks about its "connected-car" project, which conjures up images of drivers rebooting on the motorway. But the direction is clear. In future, most people in rich countries will be "always on", and will connect to the internet through something other than a PC.

The other, and related, big idea concerns what some vendors call "the digital home" and others the "e-home". This year's CES was full of mock homes in which the toaster, the refrigerator and the oven talk wirelessly to the computer, where toilet seats warm up at appropriate times and the front door can be unlocked remotely through the internet by the owner on his business trip abroad.

More than anything, however, the e-home is about digital entertainment, based on the premise that all media are currently migrating from analogue to digital form. This is happening fastest in photography. In America, by 2004 digital cameras were already outselling film-based ones, and in the rest of the world the crossover was expected to happen in 2005, according to IDC. Already lots of digital pictures are being created that need to be stored and shared.

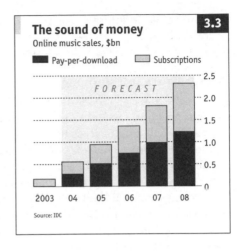

The sound of money   3.3
Online music sales, $bn
■ Pay-per-download   ☐ Subscriptions

FORECAST

2003 04 05 06 07 08

Source: IDC

Another promising medium is music, which is already digital (on CDs), but which is also increasingly being sold online, either through downloads of songs or subscriptions to online libraries (see Chart 3.3). This has already led to the revival of consumer brands such as Apple, with its hugely successful iTunes music store, and has recently attracted competition from Sony, Wal-Mart and Virgin as well as from Apple's old nemesis, Microsoft. Films and television are also moving online, albeit more slowly.

These trends raise new complexity issues. The first is the challenge of connecting all the devices in the home – the PC, the camera, the game consoles, the stereo speakers, the TV sets and even the electronic picture frames – through a wireless network, so that they can share all these digital media without too much hassle. This is crucial because, according to Ted Schadler at Forrester Research, consumers are demanding "experience liberation". In other words, they will not buy music or other media if they fear that they can only "experience" these things while sitting in front of their computer screen.

Paul Otellini, the second-in-command at Intel, expressed the challenge more poetically when he spoke at the CES. Intel and its partners, he promised, will not only Wi-Fi the home (because otherwise the tangle of cables would be offputting); they will also Veri-Fi (because everything must be totally secure), Hi-Fi (because the quality of sound and video must be good), Ampli-Fi (because the experience should reach into the garden, the garage and the basement), and of course Simpli-Fi. Mr Otellini emphasised this last point: "We need to make this dirt-simple, at ten feet, not at two feet." That is because people will no longer be sitting two feet away from a computer screen with a keyboard, but ten feet away from something or other with a remote control.

The seams are still showing

But simply making the home a perfect communications hub is not enough. According to John O'Rourke, Microsoft's director of consumer strategy, people want access to their media at all times, including when they are travelling. Gadgets must therefore know how to forward a phone call, a song or *Finding Nemo* automatically from the living room to the car. Microsoft calls this "seamless" computing; other vendors call it "pervasive" or "ubiquitous". When Mr O'Rourke recently demonstrated some of Microsoft's efforts in seamlessness at an event in Silicon Valley, at one point the Windows system that was projected on to the big screen displayed a message that it had malfunctioned and was shutting down. That seemed to ring a bell with the audience.

All this will make technology even more complex, because broadband needs to work reliably both at great distances (to connect to the internet and when roaming) and at short distances (to connect gadgets within the home). In respect of the first, for the moment the best efforts of gadget vendors such as HP and Motorola allow Wi-Fi networks within the home or office to link up with mobile-phone networks on the road. Hopes are also high for a new wireless technology called WiMax, which has a range of 31 miles, or 50km (compared with Wi-Fi's 100 yards, or 90 metres) and could therefore blanket entire cities with connectivity.

Ironically, connecting gadgets at short range in a user-friendly way could prove trickier. Even today, home networking with cables and PCs and printers is not for the faint-hearted or the over-25s. "Most consumers don't have true networks at home; they're only divvying up their internet access," says Kurt Scherf at Parks Associates, a consumer-technology consultancy. As soon as the network becomes wireless and the "nodes" include DVD players, TV and audio sets, the task becomes daunting. When Walter Mossberg, a reviewer of consumer gadgets, recently tried to connect his PC to his stereo through a fancy wireless device called Roku SoundBridge, the thing asked him whether his password was in ASCII or in Hex. Mr Mossberg, stuck for an answer, abandoned the experiment.

Help may be on the way in the form of "ultrawideband", another promising wireless technology that will connect devices over short distances at blazing speeds. However, even once ultrawideband becomes available, a lot else needs to happen before setting up an e-home becomes simple. The computing, networking and consumer-electronics industries have to agree on standards and communication protocols and on compatible copyright software. The challenge is compounded by consumers' buying habits. At present, most vendors are hawking a "highly architected, my-boxes-only strategy," says Forrester's Mr Schadler, but "nobody buys technology this way, all at once, with benefits delivered in some distant future". Instead, he says, consumers have budgets and "add home network functions one cheap device at a time". Only Apple, with gadgets such as its $129 AirPort Express, a cute little thing that plugs unobtrusively into a power socket and delivers iTunes from a PC to the stereo, gets that point, says Mr Schadler.

For other vendors, this could prove deadly. If they hawk complex products to consumers, the expenses of maintaining their own support hotlines (one customer call costs them about $30) will eat into their profits, and customers may end up angry at the brand anyway. Instead, as with every other consumer technology in history, says Parks's Mr Scherf, the digital home "must become invisible to the consumer" in order to succeed. So what should the consumer see?

The mom test

A geek's benchmark for true simplicity

"WITH E-MAIL, it wasn't till my mom could use it that it became ubiquitous. The real test is always the mom test," says Brad Treat, the boss of SightSpeed, an internet video company. "If my mother flips over to some Skype thing …," begins Michael Powell, America's media and telecoms regulator, answering a question about internet telephony. "If my mother is going to use it …," starts Ray Lane, a venture capitalist, asked whether this or that technology has a future.

Mothers come up surprisingly often in Silicon Valley conversations. Whether that is because of their unequalled wisdom, because the IT industry is full of males who are too caught up with technology to have met many women other than their mothers, or because of a misogynist streak that suspects women of a certain age to be diehard analogues is a moot point.

Grandmothers, sisters, teenage daughters and other female kin also have their place. Mr Lane, for instance, not only believes in the mom test but also has a "sister theory" to explain market inertia. This is mainly because he has a sister who spent a long career as an executive with an American airline, where she "fought every technological change over 30 years, even though she couldn't say why".

Mom, however, is invoked most – if not necessarily heeded. According to an industry legend, Steve Ballmer, now the boss of Microsoft, conducted a mom test before the launch of Windows 95, using his own mother as the guinea pig. When she had finished trying it out, Ms Ballmer asked, "How do I turn it off?" Her son, somewhat irked, pointed to the start button. "You go to the start button to stop?" asked his mother, quite perplexed. But today, several versions of Windows later, that is still how it is done.

Metaphorically speaking

What's the use of all that electronic information if you can't get at it?

THE TWO BIGGEST CONSUMER-TECHNOLOGY successes of recent times are a white page and a wheel. The white page belongs to Google, the world's most popular search engine; the wheel to Apple's iPod, the world's most popular portable music player with a hard disk. Both form part of so-called "interfaces" – metaphorical gateways through which humans enter and navigate around a technology. Both are also picture-book examples of simplicity concealing complexity underneath.

The white page is said to have come about as follows. In its early days, Google kept receiving strange anonymous e-mails containing only the number 53. Sometimes they stopped coming, then they started again. Eventually, one of Google's geniuses figured out that the e-mails arrived whenever Google had made changes to its web home page that expanded its word count beyond 53. The anonymous adviser was telling Google to keep down the clutter (although why he picked 53 as the cut-off point remains a mystery). In August 2004, Google made the biggest stockmarket debut of any technology firm in history. The current word count on google.com is 27.

As for the iPod, "It is successful because it's simple," says Paul Mercer, the brainfather of its interface and the founder of Iventor, a technology-design firm. "It does few things, but some subtle things, and it is fluid." The simplicity comes from the wheel itself; the subtlety comes from features such as the acceleration built into the wheel, so that it seems to sense whether the user wants to scroll through songs slowly or fast. The genius lies in what is absent – there is no "fast-scroll" button. Instead, says Mr Mercer, the "technology materialises only when needed", and thus "seems to intuit" the user's intention.

Google and the iPod are successful because each rescues consumers from a particular black hole of complexity. Google does it by putting a white page on top of the googol (the number 1, followed by 100 zeros) of potential web pages. The iPod does it by letting music lovers, in effect, carry all of their CDs with them in their pocket. Both solutions require an enormous technological apparatus behind the scenes. Google is said to operate some 100,000 servers. And Apple had to configure the iPod so that it automatically and fluently talks to iTunes, the music application that runs on users' PCs. Transferring songs from the PC to the iPod now requires nothing more than plugging in a single cable. (Both companies, incidentally, are notoriously secretive and refused to be interviewed for this article.)

More flops than hits

Perhaps the most startling thing about Google and the iPod, however, is the fact that they stand out so much. There are very few other recent examples of interfaces that have opened up entirely new avenues for technology to change human behaviour. Yet breakthroughs on this scale are needed if technology vendors are to see their visions come true. Those visions, remember, assume that people will increasingly connect to the internet through devices other than the PC. These gadgets will either have smaller screens, as with iPods, mobile phones or watches, or larger and more remote ones, as with TV sets or even, perish the thought, car windscreens.

Small screens require simplicity for two reasons, says Mr Mercer. One is the "lack of real estate", ie, very restricted space, meaning that not much fits on to the screen at one time. The other is that the method of input is different, because there is either only a tiny keyboard or none at all. Mary Czerwinski, a cognitive psychologist at Microsoft who calls herself the "visualisation and interaction boss", has also found big gender differences. For whatever reason, women struggle with small screens, whereas men do almost as well on them as on PC monitors.

Large screens, for their part, require simplicity because they tend to be further away than a PC monitor and operated by a remote control, or because of the context in which they might be used. "Simplicity is a must-have when you're driving," says Jack Breese, Microsoft's research director.

Even for the traditional PC, however, a new interface is needed. The present "metaphor", in designer-speak, of a desktop surface was Apple's key commercial breakthrough that launched the PC era in 1984. This broad metaphor also lent itself to sub-metaphors, including object-icons such as a rubbish bin (also the work of Mr Mercer when he worked at Apple in the 1980s), folders and files. Microsoft eventually copied these metaphors and brought them to the mass market, thus helping to make millions of computer users more productive.

But now that the internet era, in which everything is connected, is taking over from the PC era, in which computers were mostly isolated, these old metaphors are becoming increasingly redundant. PCs are turning into crowded repositories of family photographs, songs and e-mails alongside word documents and spreadsheets, and point to locations on their own hard disks as well as to computers far away. This is too much to keep track of on one desktop. "Making everything visible is great when you have only 20 things," writes Mr Norman in *The Invisible Computer*. "When you have 20,000, it only adds to the confusion. Show everything at once, and the result is chaos. Don't show everything, and stuff gets lost."

The desktop metaphor is collapsing under the weight of data overload, says Tim Brown, the boss of IDEO, a design firm in Silicon Valley. "Browsing in the old sense of the word becomes pointless," he says, and "filtering becomes crucial." This applies both to items that are stored on the user's PC and to those on the internet because, in an always-on world, the distinction becomes irrelevant.

Hence the excitement about Google. Its algorithms have so far been directed only at websites, but it plans to deploy its search technology to help people find their own documents as well. Google is currently soft-launching Gmail, a free e-mail service that offers one gigabyte of free storage. This could be a first step towards letting customers store all their data on Google's servers, where they will be easily searchable, instead of on their own PCs. In a parallel move, earlier this month Google offered free software that searches the local hard disks of PC users and displays the results much like those of a web search.

Naturally, this has struck fear into Microsoft, whose Windows system runs 94% of the world's PCs and which sees itself as the ruler of the desktop. Yet Microsoft understands the threat that data overload poses to Windows' current metaphors. Bill Gates, Microsoft's chairman and software boss, regards this interface crisis as one of the biggest challenges for his firm, alongside the security holes in Windows and, perhaps, the threat from Linux, an open-source operating system.

His plan was therefore to introduce new metaphors in the next version of Windows, code-named Longhorn. Instead of files and folders, it would use fancy new search algorithms to guide users through their PC. This technology, called WinFS (which stands either for "file system" or "future storage"), was to turn Longhorn into relational databases so that users would no longer need to remember where they put things, because the interface would automatically retrieve data for them as needed. Alas, in August 2004 Microsoft announced that Longhorn would be delayed until 2006 and that its gem, WinFS, had been dropped from it altogether. Gleefully, rivals now refer to Longhorn as either Longwait or Shorthorn.

Honey, we need to talk

Even the mockingbirds, however, cannot agree on what metaphor should replace the desktop. One favourite seems to be some kind of "personal assistant". But that may be promising too much, because what makes real-life assistants helpful is that they are able to make sense of their bosses' inchoate ramblings. In computing, says Microsoft's Mr Breese, "the holy grail of simplicity is I-just-wanna-talk-to-my-computer", so that the computer can "anticipate my needs". The technical term for this is speech recognition. "Speech makes the screen deeper," says X.D. Huang, Microsoft's expert on the subject. "Instead of a limited drop-down menu, thousands of functions can be brought to the foreground."

The only problem is that the idea is almost certainly unworkable. People confuse speech recognition with language understanding, argues Mr Norman. But to achieve language understanding, you first have to crack the problem of artificial intelligence (AI), which has eluded scientists for half a century. In fact, the challenge goes beyond AI, according to Mr Norman, and to the heart of semantics. Just think how difficult it would be to teach somebody to tie a shoelace or to fold an origami object by using words alone, without a diagram or a demonstration. "What we imagine systems of speech-understanding to be is really mind-reading," says Mr Norman. "And not just mind-reading of thoughts, but of perfect thoughts, of solutions to problems that don't yet exist." The idea that speech recognition is the key to simplicity, Mr Norman says, is therefore "just plain silly".

He concludes that the only way to achieve simplicity is to have gadgets that explicitly and proudly do less (he calls these "information appliances"). Arguably, the iPod proves him right. Its success so far stems from its relative modesty of ambition: it plays songs but does little else. In the same vein, other vendors, such as Sun Microsystems, have for years been promoting radically stripped-down devices called "network computers" or "thin clients" that do nothing but access the internet, where the real action is. Such talk horrifies firms such as Microsoft, whose financial fortunes rely on clients getting thicker so that they can sell software upgrades. But in the end the minimalists may be proved right.

Hearing voices

Plain old telephone systems are becoming redundant

SARA BAUMHOLTZ lives in Hawaii and wants to stay in close touch with her young granddaughter in Pennsylvania. So Ms Baumholtz, by inclination an "analogue", became a "digital immigrant". With the help of software from SightSpeed, a Californian firm that is at last making video-calling foolproof for ordinary humans, Ms Baumholtz now talks to, and makes faces at, the distant toddler through her PC monitor and webcam. And, because she is not using a telephone line and her broadband internet access is always on, she no longer bothers to "hang up", staying connected to Pennsylvania throughout the day. That got her thinking. "I wouldn't be surprised if I got rid of the phone," she says.

Ms Baumholtz represents the leading edge of a trend with implications that are as far-reaching as they are often underestimated. Telephony, as a stand-alone technology and as an industry, will gradually disappear. "In ten years the whole notion of a phone call or a number may be dead," says Paul Saffo at Silicon Valley's Institute for the Future. "Instant messaging (IM), audio IM, video IM – what is a call? You will click on an icon and talk, just as when you see somebody in the hallway." Or just as Ms Baumholtz does already.

Today, most people make phone calls on the "plain old telephone system" (POTS), where operators open a dedicated circuit between the callers, which can be next door or in different countries. This network consists of a set of pipes that is separate from the internet. However, voice conversations can also be sent over the internet, in the same way that e-mails travel. The caller's voice is broken into packets of digital information that are routed separately to their destination and reassembled at the other end.

In pure form, such conversations are called internet telephony. This might involve a video call between two SightSpeed customers, or a voice call between two computers that use software from Skype, a fast-growing European firm. This pure form is still rare, however, because most people still use traditional phones, which requires people calling from a PC or an internet phone to "bridge" over to the phone network. The umbrella term that includes such hybrid calls is "voice-over-internet protocol", or VOIP. This isa service offered by companies such as Vonage, a high-profile start-up in New Jersey. It allows customers to plug their old phones into an adapter that routes the call through the internet and crosses back, if necessary, to the phone network at the other end.

In the past, VOIP has not had a great reputation among consumers – if, indeed, they had heard of it at all. Today's fixed-line telephones are relatively simple devices, and mobile-phone handsets compensate for their added complexity with the convenience of mobility, so there appears to be no acute need to change the status quo. Internet telephony, by contrast, still conjures up images of geeks fiddling with their computer settings to talk to other geeks. And even if the new generation of VOIP providers, such as Vonage, really are simplifying things, they nonetheless appear at first glance to be mere substitutes for the incumbent telecoms utilities. Currently, their big selling point is not simplification but lower cost, because VOIP is much cheaper than conventional telephones, and pure internet telephony is free. That is enough reason for some consumers to make the switch (see Chart 3.4).

Companies are drawn to VOIP by its lower costs in the first place, but they also quickly discover its simplifying magic. This starts behind the scenes. Today, companies need to maintain four separate communications infrastructures. One is the data network; another is a "private branch exchange" (PBX) for external phone calls; a third is an "automatic call distributor" (ACD) to route calls internally; and the fourth is a voicemail system. By switching to VOIP, companies can ditch everything but their data network, which makes maintenance dramatically simpler for the IT staff. For instance, employees can "log on" to their phone from any cubicle or desk, whereas with POTS any office move causes expense and disruption. According to the Meta Group, a consultancy, 63% of North American companies (including giants such as Boeing and Ford) have already switched to internet telephony, either entirely or in part.

It does not take long for employees of companies with VOIP to cotton on to its many other conveniences. Today's generation of VOIP uses a technology called "session initiation protocol" (SIP), which integrates voice with other software programs, such as e-mail, instant messaging, and calendar and collaboration applications. Qualitatively, in other words, VOIP has less in common with telephones than with, say, Microsoft Outlook or Hotmail. This makes a busy executive's life simpler in several ways.

All-in-one

In a POTS world, employees can easily spend hours a week checking separate voice-mail systems in the office, at home and on their mobile phones; they also need to look out for faxes and keep an eye on their pager. To make a call, they typically go to their contacts software and then manually key a number into their phone, perhaps looking up a country code first. "Phone tag", the game played by people trying, and failing, to get hold of each other on the phone, causes frustration every day. Setting up a conference call still gets the better of many cubicle workers. Calling while travelling is messy if it involves fixed-line phones and expensive (as well as spotty) with a mobile phone.

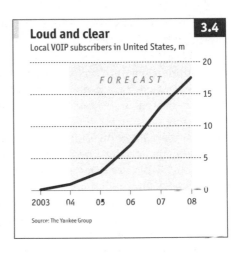

Loud and clear **3.4**

Local VOIP subscribers in United States, m

FORECAST

Source: The Yankee Group

In a VOIP world, by contrast, there is one universal in-box for voice-mails, e-mails and all other messages, which can be checked from any internet browser. Users can save their voicemails, reply to them with text, video or voice, and attach spreadsheets and presentations to their voice if appropriate. Numbers are no longer necessary, because SIP is happy with names. Initiating a call, whether to one person or many, requires only a single click on a name or an icon. Phone tag has become a thing of the past. Travelling too has ceased to be a problem: the user simply "logs on" to his phone wherever he has an internet connection.

Because that connection nowadays tends to be always on, moreover, people start changing their behaviour. Users stay permanently online with the people in their contacts application (as Ms Baumholtz does with her granddaughter), practising what Rich Tehrani, a VOIP expert, calls "ambient telephony". They will not be talking the whole time, says Mr Tehrani, but fluidly "escalating" and "de-escalating" between different levels of interaction. A conversation between several colleagues might start with a few instant text messages, then escalate to a voice or a video call, then slumber for a few hours with icons sitting at the bottom of the screen, then start again, and so on. It is rather like sharing an office or a kitchen.

Crucially, SIP also allows for social and business etiquette through a feature called "presence". For instance, a caller browsing through his contacts software may see some names (or photographs) in red, which tells him that they are busy, so he will not call them but might leave a voice or text message instead. Other contacts, such as family members, may be shown in yellow – ie, busy but available for emergencies. Others might be green, indicating that according to their calendar software they are all in the same conference call. By clicking, the caller can automatically join that conversation. Thus, says Tim Brown, the boss of IDEO, a big technology-design firm, VOIP can "make technology polite" – less intrusive, more humane and thus easier to live with.

Within the next decade, says Donald Proctor, the VOIP boss at Cisco, the world's largest maker of networking gear, VOIP could reach a tipping point, as millions of cubicle warriors, by then persuaded by the convenience of VOIP, decide to bring the simplicity of "converged" communications into their homes and disconnect their POTS utility. An obvious time for such a step might be when people move house and get fed up with spending an hour listening to their utility's muzak just to disconnect and reconnect a physical phone line.

VOIP, in other words, is today roughly where e-mail was a decade ago. Some people were predicting that e-mail would lead to world democracy, if not nirvana, whereas analogue sceptics insisted that it was just a paper-saving alternative to office memos. Then people started bringing their e-mail habits home from the office. Since then, e-mail has become radically simpler, unbound from geography and ubiquitous. It has made communicating with far-flung friends free and easy (although, arguably, it now has to defend that convenience against spam). And as it got simpler, it simplified its users' lives.

VOIP has the same potential. It may not be for everyone yet, but over the next decade, as the fiddliness of connecting to the internet – whether through the air, the power socket, the old phone jack, the cable-TV dongle, or by satellite – is resolved, that connection will increasingly be the only link needed. Communicating, by voice or any other means, will be free. Will it be simpler? Ask Ms Baumholtz.

The blood of incumbents

Stand by for a spot of creative destruction

O N THE RECORD, any top executive in the IT, consumer-electronics and telecoms industries today will profess that his firm is leading the way towards simplicity. But are those claims justified? In theory, says Ray Lane, a venture capitalist, the company best placed to deliver simplicity should be Microsoft. It controls virtually all of the world's PCs and laptop computers (albeit smaller shares of mobile phones, hand-held and server computers), so if its software became simpler, every-thing else would too. The bitter irony, says Mr Lane, is that Microsoft is one of the least likely companies to make breakthroughs in simplifica-tion. "It cannot cannibalise itself," says Mr Lane. "It faces the dilemma."

The dilemma? These days, whenever anybody in the IT industry mentions that word, it is instantly understood to refer to *The Innovator's Dilemma* (Harvard Business School Press, 1997), a book by Clayton Christensen, a professor at Harvard Business School, who has since fol-lowed it up with a sequel, *The Innovator's Solution* (Harvard Business School Press, 2003). In a nutshell, the dilemma is this: firms that succeed in one generation of innovation almost inevitably become hamstrung by their own success and thus doomed to lose out in the next wave of innovation. Just as they "disrupted" the previous era's leaders, they are in turn disrupted by the pioneers of the next era.

To explain how this happens, Mr Christensen distinguishes between two basic types of innovation. The first is "sustaining" inno-vation. This is the sort that incumbent firms are engaged in to sell ever better, and ever more profitable, products to their most attractive and demanding customers. An example might be Microsoft adding more features to Word, Excel and PowerPoint. If challenged by upstarts, incumbents almost always prevail. At some point, however, the tech-nology goes into "overshoot", where users no longer have an appetite for additional bells and whistles, and sustaining innovation leads to numbing complexity.

At this point, according to Mr Christensen, the second, "disruptive", type of innovation becomes possible. Disruptive technologies target the least demanding customers in the current market, or even entirely new markets of "non-consumers", by offering something simpler, or cheaper, or both. An example of a disruptive technology that is cheaper but not necessarily simpler is Linux, an open-source operating system that is taking market share from Unix and Windows. And in disruptive-technology battles, Mr Christensen argues, newcomers to the industry almost invariably "crush the incumbents".

One reason is an asymmetry in financial incentives. A disrupter might look at a million non-consumers and see a huge opportunity, whereas the incumbent sees a drop in the ocean. Initially, moreover, the incumbent will find being disrupted very pleasant, because the customers that defect first are likely to be the unprofitable ones. As its own profit margins improve, the incumbent will be tempted to ignore the competition. The disrupter now makes its own sustaining innovations until its technology becomes "good enough" to poach the original market, at which point the incumbent is gored.

Another reason why newcomers prevail is a cultural malaise that infects incumbents. Big, successful companies are organised into product divisions, whose managers will keep a close eye on their known rivals' offerings to ensure that their own products retain their edge. The disrupters, however, do not care about products. They observe real people, and specifically non-consumers, to see what jobs they are trying to get done. Today, for instance, Microsoft might take comfort from the fact that Excel has more features than any other spreadsheet, whereas a potential disrupter might note that people are driven to despair when trying to transfer files from an old to a new computer.

Thus, technology has historically advanced in waves of disruption. The original Bell (the ancestor of modern giants such as AT&T, SBC, Verizon and Lucent) began in the late 19th century as a disrupter to Western Union. At the time telephone signals could travel for only three miles, whereas Western Union's telegraphs could communicate over long distances. Bell started by targeting the local market, but eventually improved its technology and entered the long-distance market, rendering telegraphs obsolete.

Sony became famous as a serial disrupter, starting in the 1950s, when its transistor radios skewered the radio standard of the day, vacuum tubes, and that technology's incumbent, RCA. In the 1970s and 1980s, Xerox was the incumbent in photocopiers, rebuffing "sustaining" challenges by IBM and Kodak to make better copiers for the top end of the market, before succumbing to the disruption from simple and cheap table-top copiers from Canon. IBM in turn was the incumbent in mainframes and parried "sustaining" attacks from General Electric, RCA and AT&T, until mainframes were disrupted by PCs and firms such as Microsoft, Intel and Dell. And so on.

Who's next?

Where does that leave the IT, consumer-electronics and telecoms industries today? Many of their current products have far overshot the needs of businesses and consumers, yet failed to help them to get essential jobs done. Moreover, billions of analogues will eventually become digital immigrants, whether for fear of social isolation in rich countries or, in developing countries such as India and China, because they will be able to afford to. These current non-consumers are technology's next frontier.

For corporate buyers of IT, it has become clear in the past few years that their ability to "get jobs done" no longer has much to do with the power and complexity of their computers. Instead, they are increasingly finding that the simplest way to keep track of customers, bills, inventories and so forth is to rent such services for a monthly fee. This suggests that application service providers (ASPs) such as Salesforce.com, in their business models as well as in their technologies, could become disruptive simplifiers at the expense of today's enterprise-software giants.

For consumers, it is increasingly clear that coping with information overload is a big "job to be done". Google has already acted on that observation by disrupting various old-fashioned owners of directories, such as Yellow Pages. Having moved well ahead with its own sustaining improvements, Google (or a firm like it) stands a chance of becoming a disruptive simplifier at the expense of incumbents such as Microsoft, which does not let consumers store information by content across all applications, making it harder to get at. In telecommunications, mobile phones have for years been disrupting the incumbent fixed-line providers, but now they themselves are in danger of overshooting. Capgemini, a consultancy, has found that most mobile-phone operators vastly overestimate the importance that customers place on premium services, while equally vastly underestimating the importance of simplicity, both in handsets and in pricing plans. This is opening the door to disrupters such as Comviq, in Sweden, which has taken 39% of market share away from the incumbent, Telia, by offering half as many handset features and radically simpler pricing plans.

Wireless and fixed-line telephone companies may simultaneously become vulnerable to new providers of internet telephony or VOIP, such as Skype and Vonage, or networking companies such as Cisco (especially once fast, wireless internet access has become ubiquitous and totally reliable). The disruption could be especially severe if the upstarts not only make calling dirt-cheap or free, but also find ways to help consumers with jobs such as simplifying their communications as a whole or meeting their needs for privacy.

For incumbents this ought to be reason for paranoia, but it need not spell doom. If they play their cards right, they too can take part in the game of disruption – as AT&T, for instance, is trying to do by withdrawing from the residential telephone market at the same time as vigorously marketing its own VOIP service. The key will be to aim for simplicity and affordability.

Everybody else, meanwhile, has cause for optimism. A lot of things that are complex today will get simpler in the coming years. Like other technologies in history, IT and telecommunications seem destined gradually to recede into the background of human activity, leaving more time and energy to get on with the infinite complexities of business, and of life in general.

The material on pages 78–110 first appeared in a survey in *The Economist* in October 2004.

4
A WORLD OF WORK

A world of work

The global deployment of work has its critics, but it holds huge opportunities for rich and poor countries alike

O N A TECHNOLOGY CAMPUS off the bustle of the Hosur Road in Electronics City, Bangalore, engineers are fiddling with the innards of a 65-inch television, destined for American shops in 2006. The boffins in the white lab coats work for Wipro, an Indian technology company. Wipro has a research-and-development contract with a firm called Brillian, an American company based half a world away in Tempe, Arizona. Brillian's expertise is in display technology. Wipro's job is to put together the bits that will turn Brillian's technology into a top-end TV.

Wipro is sourcing the television's bits and pieces from companies in America, Japan, Taiwan and South Korea. After design and testing, assembly will pass to a specialist contract manufacturer, such as Flextronics or Solectron. The buyer of the finished television might use a credit card administered from Kuala Lumpur, Malaysia. After-sales service might be provided by a polite young Indian call-centre agent, trained in stress management and taught how to aspirate her Ps the American way.

A few years ago, the combination of technology and management know-how that makes this global network of relationships possible would have been celebrated as a wonder of the new economy. Today, the reaction tends to be less exuberant. The same forces of globalisation that pushed Flextronics into China and its share price into the stratosphere in the 1990s are now blamed for the relentless export of manufacturing jobs from rich to poorer countries. Brillian's use of Indian engineers is no longer seen as a sign of the admirable flexibility of a fast-growing tech firm, but as a depressing commentary on the West's declining competitiveness in engineering skills. The fibre-optic cable running between America and India that used to be hailed as futuristic transport for the digital economy is now seen as a giant pipe down which jobs are disappearing as fast as America's greedy and unpatriotic bosses can shovel them.

These anxieties have crystallised into a perceived threat called "outsourcing", a shorthand for the process by which good jobs in America, Britain or Germany become much lower-paying jobs in India, China or

Mexico. Politicians decry outsourcing and the bosses they blame for perpetrating it. The same media that greeted the rise of the new economy in the 1990s now mourn the jobs that supposedly migrate from rich countries to less developed ones.

Forrester, an American research firm, has estimated these future casualties down to the last poor soul. By 2015, America is expected to have lost 74,642 legal jobs to poorer countries, and Europe will have 118,712 fewer computer professionals. As Amar Bhide of Columbia University comments drily, "Graphs from a few years ago that used to predict explosive growth in e-commerce have apparently been relabelled to show hyperbolic increases in the migration of professional jobs."

Amid all this clamour, some of the vocabulary has become mixed up. Properly speaking, outsourcing means that companies hand work they used to perform in-house to outside firms. For example, Brillian is outsourcing the manufacture of its televisions to Flextronics or Solectron. Where that work should be done involves a separate decision. Flextronics might assemble bits of its televisions in Asia but put together the final products close to its customers in America. If it does, it will have moved part of its manufacturing "offshore". Not all offshore production is outsourced, however: Brillian might one day open its own "captive" research-and-development facility in Bangalore, for instance.

What agitates worriers in the West is the movement of work abroad, regardless of whether it is then outsourced or performed in-house. But the reality is more complicated than they acknowledge.

A well-established model

The age of mass mechanisation began with the rise of large, integrated assembly lines, such as the one Henry Ford built in 1913 at Dearborn, Michigan, to make the Model T. Over the course of the 20th century, companies reorganised industrial production into ever more intricate layers of designers, subcontractors, assemblers and logistics specialists, but by and large companies have mostly continued to manufacture close to where their goods are consumed. They have then grown internationally by producing overseas, for new customers, the same goods they produce and sell to their customers at home: 87% of foreign direct investment is made in search of local markets, according to McKinsey, a consultancy. Products and brands have become global, but production has not.

Conversely, white-collar work continues to be produced in the same way that Ford produced the Model T: at home and in-house. Bruce

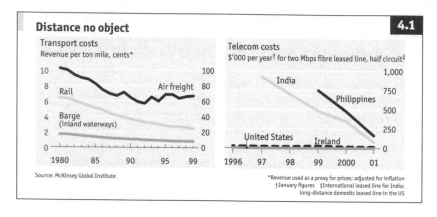

Distance no object `4.1`

Transport costs
Revenue per ton mile, cents*

Rail
Air freight
Barge
(inland waterways)

Source: McKinsey Global Institute

Telecom costs
$'000 per year† for two Mbps fibre leased line, half circuit‡

India
Philippines
United States
Ireland

*Revenue used as a proxy for prices; adjusted for inflation
†January figures ‡International leased line for India;
long-distance domestic leased line in the US

Harreld, the head of strategy at IBM, reckons that the world's companies between them spend about $19 trillion each year on sales, general and administrative expenses. Only $1.4 trillion-worth of this, says Mr Harreld, has been outsourced to other firms.

Brillian obtains both the goods and the services it needs to put together its televisions from outsiders all over the world, which means each bit of work goes to whatever company or country is best suited to it. This opens up huge opportunities. Diana Farrell, the head of McKinsey's Global Institute, thinks that by reorganising production intelligently, a multinational firm can hope to lower its costs by as much as 50–70%.

Such reorganisation takes two main forms. First, thanks to the spread of the internet, along with cheap and abundant telecommunications bandwidth, businesses are able to hand over more white-collar work to specialist outside suppliers, in the same way as manufacturers are doing already. A growing number of specialists offer, say, corporate human-resources services, credit-card processing, debt collection or information-technology work.

Second, as transport costs fall, globalisation is beginning to separate the geography of production and consumption, with firms producing goods and services in one country and shipping them to their customers in another. Over the past decade, countries such as Mexico, Brazil, the Czech Republic and, most notably, China have emerged as important manufacturing hubs for televisions, cars, computers and other goods which are then consumed in America, Japan and Europe. Such offshore production is central to the strategies of some of the world's most powerful businesses, including Wal-Mart and Dell.

Over the next decade, Russia, China and particularly India will emerge as important hubs for producing services such as software engineering, insurance underwriting and market research. These services will be consumed at the other end of a fibre-optic cable in America, Japan and Europe. Just as Dell and Wal-Mart are obtaining manufactured goods from low-cost countries, companies such as Wipro, TCS and Infosys, for instance, are already providing IT services from low-cost India.

As businesses take advantage of declining shipping costs, abundant and cheap telecommunications bandwidth and the open standards of the internet, the reorganisation of work in each of these areas is likely to advance rapidly. IBM's figures suggest that companies have so far outsourced less than 8% of their administrative office work. Privately, some big companies say that they could outsource half or more of all the work they currently do in-house.

Rich-country manufacturers have already invested hundreds of billions of dollars in building factories in China to make clothes, toys, computers and consumer goods. In the next few years, they may invest hundreds of billions more to shift the production of cars, chemicals, plastics, medical equipment and industrial goods. Yet the globalisation of white-collar work has only just begun.

A study by McKinsey looked at possible shifts in global employment patterns in various service industries, including software engineering, banking and IT services. Between them, these three industries employ more than 20m workers worldwide. The supply of IT services is the most global. Already, 16% of all the work done by the world's IT-services industry is carried out remotely, away from where these services are consumed, says McKinsey. In the software industry the proportion is 6%. The supply of banking services is the least global, with less than 1% delivered remotely.

McKinsey reckons that in each of these industries, perhaps as much as half of the work could be moved abroad. But even a much smaller volume would represent a huge shift in the way that work in these industries is organised. There may be just as much potential in insurance, market research, legal services and other industries.

Outsourcing inspires more fear about jobs than hope about growth. But the agents of change are the same as those that brought about the 1990s boom. New-economy communications and computer technologies are combining with globalisation to bring down costs, lift profits and boost growth. This section will try to restore some of the hope.

Men and machines

Technology and economics have already revolutionised manufacturing. White-collar work will be next

THE INDUSTRIAL COMPLEX that Henry Ford built on the banks of the Rouge River in Dearborn, Michigan, was a wonder of the new age of mass production. Into one end of the plant went iron ore, coal, sand and rubber, brought in by railway and on Great Lakes steamships. Out of the other end rolled Model T Fords. By 1927, there had been 15m of them. At that stage, Dearborn was handling every step of the car's production, from rolling steel to making springs, axles and car bodies, and casting engine blocks and cylinder heads. The plant even had its own glass factory.

Ford built the Dearborn plant around the labour-saving properties of machines. Automation lowers production costs, which bolsters profits. Companies spend these profits on improving what they sell, and on building more labour-saving machines. As technology advances, these improvements make products more complex. To the basic design, modern carmakers add heated seats, air conditioning, guidance and entertainment systems, computer chips that regulate engine performance, and many other gadgets to please their customers. It took 700 parts to make the Model T. Modern cars pack many more into their radios alone.

As industries advance, manufacturers manage the growing complexity of their products by outsourcing: they share the work of making them with others. This enables each company in the production chain to specialise in part of the complicated task. The car industry, for instance, relies on parts companies that make nothing but electrical systems, brakes or transmissions. These parts companies, in turn, depend on the work of other suppliers to make individual components. At each level of production, outsourcing divides up growing complexity into more manageable pieces.

In the office, the tool used to mechanise work is the computer. Computers automate paperwork and hence the flow of information. Companies that sell information products, such as banks and insurance firms, employ computers to automate production. And all companies use computers to automate the administrative work needed to maintain their organisations: keeping their books in good order, complying with rules and regulations, recruiting, training and looking after their employees, managing offices, dealing with company travel and so on.

Bells and whistles

Like assembly-line machinery, computers save labour, bring down costs and raise profits. Banks and insurance companies have used some of these profits to add bells and whistles to their products, making them more complex. Banks that used to provide basic mortgages now sell fixed loans and floaters, caps, collars, locks and other financial exotica to befuddled home-buyers. Credit card companies offer loyalty programmes, membership rewards and cash-back deals. Insurance firms tailor car and life insurance to fit their customers' appetite for risk.

Corporate administrative work has also become more complicated. The demands of securities regulators and investors for financial information have expanded with the capacity of firms to supply it. IBM's annual report for 1964 contains a scant half-dozen pages of financial information; its most recent one includes 40 pages of financial statements and accounting notes. The more services that corporate HR departments provide to employees, the more employees expect. Evermore prescriptive accounting and audit rules proliferate as fast as accounting departments can automate the work of complying with them.

The spread of computers through companies has added a third layer of complexity: the task of managing the information systems themselves. The work of company IT departments is particularly complicated at older and larger firms that have bought different sorts of computer systems at different times. The core processing systems of insurance companies, airlines and banks, for instance, are built on a mainframe-computer technology that celebrated its 40th anniversary in 2004. Companies have added extra systems as they have sold new products, grown abroad or acquired competitors. Most IT departments at most large companies spend most of their time simply fighting to keep this tangle of systems going.

In all three areas of white-collar work, companies are struggling to manage growing complexity. The chief reason for the recent recession in corporate IT spending is that the IT industry's customers are no longer able to absorb new technologies, thinks IBM's Mr Harreld. Entangled in new products and the computer systems that support them, banks cannot even do something as basic as ensuring that customers who

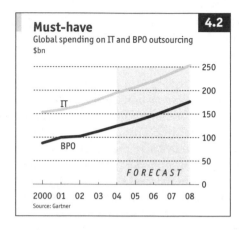

Must-have `4.2`
Global spending on IT and BPO outsourcing
$bn

```
                                    250
                                    200
   IT                               150
                                    100
   BPO
                                    50
           FORECAST
                                    0
2000 01  02  03  04  05  06  07  08
Source: Gartner
```

asked one department not to send junk mail do not receive it from another. "If a bank was making cars, every tenth car would come out without a steering wheel," says Myles Wright of Booz Allen Hamilton, a consultancy.

Just as in manufacturing, the solution to the growing complexity of white-collar work is to do less of it in-house. Some companies have outsourced the work of their IT departments, from managing the physical hardware to maintaining and developing business software and managing corporate computer networks. Up to half the world's biggest companies have outsourced some IT work, reckons IBM.

As well as outsourcing their business systems, some companies are doing the same with the workers who operate them. This is called business-process outsourcing (BPO). First Data Corporation (FDC), for instance, will handle some or all of the administrative work involved in running a credit-card business, from dealing with applications to authorising credit limits, processing transactions, issuing cards and providing customer service. Few bank customers will have heard of the company, yet FDC employs nearly 30,000 people, who administer 417m credit-card accounts for 1,400 card issuers.

Likewise, companies are outsourcing chunks of administrative work and their supporting systems. Accounting departments are farming out tasks such as processing invoices and collecting payments from debtors. HR departments have shed payroll work. ADP, a payroll-outsourcing company, pays one in six private-sector workers in America. Increasingly, big companies are handing over entire HR departments and the systems that support them to outside specialists such as Hewitt, Accenture and Convergys, says Duncan Harwood of PricewaterhouseCoopers.

One way for manufacturers to manage growing complexity is to adopt common standards. Carmakers, for instance, have reworked their manufacturing processes so they can assemble different car models from the same production "platform", with several cars sharing a number of parts. This allows parts companies to specialise more and produce fewer parts in larger numbers.

Eventually the organisation of car manufacturing may begin to resemble production in the consumer-electronics industry, where the adoption of industry-wide standards (along with de facto standards, such as the Intel microprocessor) has enabled suppliers to become highly specialised. Companies such as Flextronics and Selectron now offer outsourced manufacturing platforms for whole categories of consumer electronics. All the branded makers have to do is handle the logistics, badge the goods and send them off to the shops.

A similar platform-production system is emerging in white-collar work. A few popular business-software packages sold by companies such as SAP, a German software firm, and PeopleSoft, an American one, are now offering standard ways of organising and delivering administrative office work. When companies outsource HR departments, specialists such as Hewitt and Accenture add them to their HR-services production platform. Convergys, for instance, claims to be the world's largest operator of SAP's HR software. FDC, for its part, has built a production platform that offers credit-card services.

Thanks to the internet's open standards, extreme specialisation is now emerging in outsourced business services, just as it did earlier in consumer electronics. Next door to a Safeway supermarket on the Edgware Road in London, a group of British accountants and tax experts has built a business service called GlobalExpense that handles employees' expenses over the internet. Employees of its customer companies log on to the GlobalExpense website, record their expenses on standard forms and put their receipts in the mail. GlobalExpense checks the receipts, pays the expenses and throws in a few extras such as related tax work and information on whom the company's employees are wining and dining.

This year GlobalExpense will pay out £60m-worth of employee expenses, which probably makes it the biggest expense-payer in Britain. With a large, flexible pool of foreign students in London to draw on, the company says it can handle expense claims and receipts from anywhere in the world.

And so to Bangalore

In the late 1980s and early 1990s, as transport and communications costs fell and logistics technology improved, rich-country manufacturers began moving production to cheaper nearby countries. American

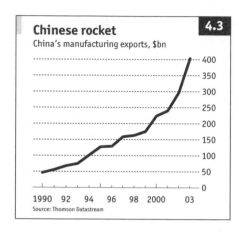

Chinese rocket `4.3`

China's manufacturing exports, $bn

400
350
300
250
200
150
100
50
0

1990 92   94   96   98   2000   03

Source: Thomson Datastream

carmakers and consumer-electronics firms started manufacturing in Mexico; European makers went to the Czech Republic, Slovakia and Poland; and Japanese, Taiwanese and Korean firms moved to China. By the late 1990s, European manufacturers such as Philips, Siemens and Nokia, and American ones such as GE and Motorola, were moving further afield, to China. American imports from China rose from $66 billion in 1997 to $163 billion last year. By one estimate, foreign companies opened 60,000 factories in China between 2000 and 2003. The country's exports rocketed (see Chart 4.3).

In the same way, with the cost of telecommunications bandwidth falling, some firms in rich countries, mostly in America and Britain, began moving some of their business services abroad, so far mostly to India. IT-service companies such as IBM, EDS and Accenture have hired thousands of Indian software engineers to carry out work previously done near their customers in rich countries. An Indian GE subsidiary called GECIS handles administrative processing work for the firm's financial businesses. NASSCOM, the Indian IT-industry lobby, has high hopes for these young export industries. By 2008, it thinks, they will employ over 4m Indians, generating up to $80 billion-worth of sales.

Firms may choose to outsource work when they move it abroad, and they may not. But actually moving particular operations abroad is more akin to introducing labour-saving machinery than to outsourcing in the sense of improving the management of complexity. It brings down the cost of production, mostly by making use of cheaper employees.

Sometimes companies even change their technology when they move abroad, making their production less automated so they capture more benefits from lower labour costs. For example, some big carmakers are reconfiguring their production to use more manual work in their Chinese factories than they do elsewhere, says Hal Sirkin of the Boston Consulting Group. Wipro Spectramind, an Indian firm, recently moved work for an American company to India. This work involved 100 people, each of whom cost the firm $6,000 in software-licence fees. The American company had been trying to write software to automate some of this work and reduce its licence-fee payments. Wipro scrapped the software project, hired 110 Indians and still did the work more cheaply.

Once work has moved abroad, however, it joins the same cycle of automation and innovation that pushes technology forward everywhere. Optical-character-recognition software is automating the work of Indian data-entry workers. Electronic airline tickets are eliminating some of the ticket-reconciliation work airlines carry out in India. Eventually, natural-language speech recognition is likely to automate some of the call-centre work that is currently going to India, says Steve Rolls of Convergys, the world's largest call-centre operator.

All this helps to promote outsourcing and the building of production platforms in India. GE is selling GECIS, its Indian financial-services administrator, and Citibank, Deutsche Bank and others have disposed of some of their Indian IT operations. Thanks to the growth of these newly independent firms, along with the rapid development of domestic Indian competitors, such as Wipro and Infosys, companies will increasingly be able to outsource work when they move it.

Dashing white collars

Manufacturing has already gone a long way down the road of outsourcing and globalisation, but there are now fears that white-collar work will be reorganised much more quickly and disruptively, thanks to the spread of the internet, plummeting telecommunications costs and the realisation that the machines used by millions of expensive white-collar workers in the West could be plugged in anywhere.

Manufacturers' shipping costs have declined more slowly than the telecommunications costs of providers of remote services. The logistics of shipping goods over long distances remain complicated and inexact. For example, the v6 car engines that Toyota sends from Nagoya in Japan to Chicago take anywhere between 25 and 37 days to arrive, forcing the car company to hold costly stocks. The movement of white-collar work, on the other hand, is subject to no physical constraints. Communications are instant and their cost is declining rapidly towards becoming free.

Yet powerful barriers to moving white-collar work remain. When work moves out of a company, the firm negotiates a commercial agreement to buy it from a supplier. For manufacturers, this is straightforward: they take delivery, inspect the goods and pay their suppliers. Supplying a service, by contrast, is a continuous process. The outsourcing industry has evolved legal contracts in which suppliers bind themselves to deliver promised levels of service. There has been much legal innovation around these contracts, not all of which has been satisfactory (see pages 123-4). The upshot is that it still takes trust and cross-cultural understanding to achieve a good working relationship. Moving a company's IT department to India is likely to put such understanding to the test.

The other big barrier is that, despite the spread of business machines, white-collar work still tends to be much less structured and rule-bound than work done on the shop floor. Unstructured work is hard to perform over long distances: without guidance, workers are apt to lose their way. The most likely outcome is that would-be outsourcers will proceed in two steps. First they will hand IT services, administrative tasks and other white-collar work to trusted specialist suppliers close to home. But once those suppliers have added structure, rules and standards, the outsourcers will move the work abroad.

A desperate embrace

Companies do not always outsource for the best of reasons

IN 2001 AND 2002, KPN, a Dutch telecoms carrier, signed several long-term deals to outsource 80% of the work done by its IT department to Atos Origin, a European provider of IT services. Three years later, both parties are still putting a lot of effort into reworking these contracts. It shows that not all decisions to outsource are straightforward and problem-free.

In 2001, KPN, like most telecoms firms, was in desperate trouble: having run up huge debts as it expanded during the telecoms bubble, it was close to bankruptcy. Atos Origin said it could help, and not just with the IT. In return for a guarantee from KPN to buy about €300m-worth of IT from it every year for the next six years, Atos Origin paid KPN €206m up front for the IT assets that the telecoms firm had handed over.

But as the spread of mobile phones and digital fixed-line technology ate into KPN's sales, the firm had to make drastic cuts. Within two years its headcount had shrunk from 28,000 to 18,000. It was now less than two-thirds its size when it signed its IT deal, yet it was still bound by contract to buy the same €300m-worth of IT services a year.

Neither party, however, could easily walk away. The solution they agreed was that Atos Origin would work to transform KPN's IT systems by the end of 2006. KPN's fixed-line division, for instance, runs 779 different applications, which the company itself thinks it can shrink to 80. That should keep its IT purchases up for a while, and so avoid any immediate damage to Atos Origin's revenues. After that, hopes Atos Origin, it will have earned the right to more transformation work from its customer, thus maintaining the value of its original contract.

Whether such "transformational" agreements are the best way forward is the subject of much debate in the industry. Supporters argue that they help to align the interests of outsourcing firms with those of their customers. Critics say they are a way of landing the industry's customers with the risk that something may go wrong: the criteria for a successful transformation are sufficiently nebulous for clever lawyers to claim that they have been met, whatever the outcome.

The larger issue, however, is the way IT firms sell financial engineering along with their systems and software. Governments, for instance, are avid advocates of long-term contracts because they can spread the cost of a large IT investment over many years, making it look more manageable. So long as the industry continues to offer this sort of balance-sheet support along with the technological variety, its customers may sometimes be tempted to make the wrong decision.

The place to be

In the global market for white-collar work, India rules supreme. But others are lining up

MOST AMERICANS OR BRITONS would be hard pressed to name their national call-centre champions or top providers of IT services. In India they are like rock stars, endlessly featured in the media. All of them claim to be hiring by the thousands every month. New business models come and go. Hero bosses such as Raman Roy, chief executive of Wipro Spectramind and "father of Indian business-process outsourcing" (an industry all of six years old), have developed the same preposterous swagger adopted by erstwhile leaders of America's dotcom boom. Is India heading for a fall, too?

India's IT industry is growing at a vertiginous rate. A dozen years ago, the entire country boasted just four or five IBM mainframe computers, says Lakshmi Narayanan, the boss of Cognizant, a big Indian IT-service company. In 2003 the industry notched up sales of $16 billion, three-quarters of which went abroad, according to NASSCOM, a lobby group. By 2008, says NASSCOM, annual sales are likely to surpass $50 billion. The big firms are hiring about 1,000 graduates a month straight from Indian technical colleges.

The sales of Infosys alone, one of the top providers of IT services, grew more than eightfold in five years, to over $1 billion in the year ending in March 2004. The firm claims to run the biggest corporate training facility in the world, with 4,000 students at a time and three courses a year. The company's chairman, Narayana Murthy, says Infosys is going to expand further.

India's BPO industry is younger and smaller, but growing even faster. In 2003 its sales were $3.6 billion; by 2008 they are expected to reach $21 billion-24 billion, says NASSCOM. About 70% of the BPO industry's revenue comes from call-centres; 20% from high-volume, low-value data work, such as transcribing health-insurance claims; and the remaining 10% from higher-value information work, such as dealing with insurance claims. But the BPO industry is more fragmented than the IT business, and could change shape rapidly.

The roots of India's competitiveness in IT reach back to the late 1980s, when American firms such as Texas Instruments and Motorola came to

Bangalore for the local talent. Other American firms, such as Hewlett-Packard, American Express, Citibank and Dun & Bradstreet, followed these pioneers, setting up their own "captive" Indian IT organisations in the 1990s.

The Indian companies got their first big boost with the so-called "Y2K crisis" at the turn of the millennium. IT experts feared that because elderly software code allowed only two digits to record the year, some computer systems would read the year 2000 as 1900, causing mayhem as systems crashed. Big western IT-services companies such as IBM, Accenture and EDS ran out of engineers to check old code and subcontracted some of the work to Indian firms instead.

Once the Indians had saved the world, they set out to conquer it. Wipro, TCS, Infosys and their peers grabbed a growing share of the global giants' business. They made most inroads in the routine but costly business of maintaining business-software applications from vendors such as PeopleSoft and SAP.

As the Indian firms grew, the captive operations of foreign firms became less competitive, and most of them have now sold out. Dun & Bradstreet led the field, with its captive transforming itself into Cognizant in 1994. More recently, Citibank sold some of its Indian IT operation to an Indian financial-software specialist called Polaris. Deutsche Bank sold its captive to HCL, another Indian firm. The big western IT specialists, meanwhile, have squared up to the new, low-cost competition by hiring in India themselves. Accenture's Indian payroll shot up from 150 in 2001 to about 10,000 in 2004.

India's BPO industry also started with foreign captives. The pioneers were GE, American Express and British Airways, who all arrived in the late 1990s. These companies were joined by home-grown call-centre operators such as 24x7, vCustomer, Spectramind and Daksh. Spectramind has since been bought by Wipro, and Daksh by IBM.

These Indian firms also face competition from specialist American call-centre companies which, like the global IT firms, have been adjusting to the cheap Indian competition by taking themselves to India. By far the most successful of these foreign firms has been America's Convergys, which with a total of around 60,000 employees is the biggest call-centre operator in the world. By the end of 2005, says the company's local boss, Jaswinder Ghumman, Convergys hoped to employ 20,000 people in India. A fourth wave of BPO start-ups, many of them funded by American venture capitalists, has been experimenting with the remote delivery from India of all sorts of work, from hedge-fund administration to pre-press digital publishing.

In both the IT and the BPO industries, the leading companies in India are fighting hard to win a broader variety of work, particularly higher-value activities. EXL Service

Sustainable? 4.4

IT and BPO employment in India, '000

	2002	2003	2006*	2009*	2012*
IT	106	160	379	1,004	2,717
BPO	170	205	285	479	972

Source: NASSCOM *Forecast

carries out a broad range of insurance work for British and American firms, from finding customers to underwriting policies, administering claims, changing policies and providing customer services. The company is a licensed insurance underwriter in 45 American states, with applications for the remaining states pending. "These are very high-end jobs," says EXL Service's boss, Vikram Talwar.

The fancy stuff

In September, ICICI OneSource, an Indian BPO company which has so far concentrated on call-centre work, took a 51% stake in Pipal Research, a firm set up by former McKinsey employees to provide research services for consultants, investment bankers and company strategy departments. Mr Roy of Wipro Spectramind says that his firm is moving from basic call-centre work – helping people with forgotten passwords, for instance – to better-quality work in telesales, telemarketing and technical support. Wipro Spectramind is also spreading into accounting, insurance, procurement and product liability. "We take the raw material and convert it," says Mr Roy, his eyes gleaming. "That is our skill – to cut and polish the raw diamonds."

The top end of the market is more interesting still. Viteos, an Indian start-up, pays new MBA graduates in Bangalore $10,000 a year to administer American hedge funds, work that involves reconciling trades and valuing investments for a demanding set of customers. Shailen Gupta, who runs an offshore advisory consultancy called Renodis, has been helping one of his American customers to hire Indian PhDs to model demand planning.

The best Indian IT and BPO companies are aiming not only to lower the cost of western white-collar work, from software programming to insurance underwriting, but to improve its quality as well. Firms such as Wipro, EXL Service and WNS, a former British Airways BPO captive that won its independence in 2002, are applying the same management disciplines to the way they provide services that GE applies to its industrial businesses. Tasks are broken into modules, examined and reworked to reduce errors, improve consistency and speed things up.

In both industries, the influence in India of GE, which has applied the "six sigma" method of quality improvements to its industrial businesses for years, is pervasive. Mr Roy used to run GECIS, which was then GE's BPO captive but is now being sold. It had become "too fat and happy", according to one Indian competitor. One of the founding investors in Mr Talwar's company is Gary Wendt, the former head of GE's financial businesses. Wipro's chairman, Azim Premji, has introduced so many of GE's techniques to his company that the firm is known as India's "baby GE".

Certainly, "Wiproites" seem to share the intensity of GE's employees. Six-sigma "black belts" hurtle about Wipro's 100-acre technology campus in Bangalore, improving everything from software coding to the way the company cleans its toilets. (Among other things, this involves analysing liquid-soap availability, tissue supply and waste management, explains a serious-looking Wipro official.)

The claims of India's marketing men tend to be a little ahead of reality. Amar Bhide of Columbia University, who has spent some time in Bangalore, is sceptical. The Y2K crisis pushed "the grungiest IT work on to India's best software engineers," says Mr Bhide. "It was like asking Oxford graduates to dig ditches. It created the impression that Indians were fantastic at programming."

Still, the outline of a distinct brand of Indian competitiveness – in performing carefully defined, rules-bound, repetitive white-collar business work – appears to be taking shape. Already, the Indian IT firms, along with some of the foreign captives in India, boast the world's most impressive set of international quality certifications for software engineering.

In the longer term, India's success at winning global white-collar work will depend on two things: the supply of high-quality technical and business graduates; and, more distantly, an improvement in India's awful infrastructure.

India's most often-cited advantage is its large English-speaking population, which has helped to fuel the call-centre boom. Yet already the market for call-centre workers is tightening. Pay and staff turnover are shooting up as operators poach staff who have already undergone costly "accent neutralisation" training at rival firms. Even the best call-centre operators in India lose about half their employees each year (but then turnover in British call-centres is about 70%). One Convergys job advertisement in the *Times of India* promises to make prospective call-centre employees "a prime target of all the dons of the industry. You will be hunted down, with almost a king's ransom on your head".

No dream job

Part of the problem is that call-centre work tends not to be much fun – although Indians enjoy much better pay, relative to other local jobs, than British or American call-centre employees. At Wipro Spectramind, two "fun day" employees try to jolly the place up as rows of cubicle-farm workers use a piece of software called "retention buddy 1.3" to dissuade Americans from cancelling their internet subscriptions. Sanjay Kumar, the boss of vCustomer, one of the few remaining independent Indian call-centre companies, says the industry's growth potential may be limited. He thinks the total pool of call-centre workers is only about 2m, and awkwardly scattered across India – although that still leaves a lot of room for expansion from the current 300,000 or so.

According to official figures, India produces about 300,000 IT engineering graduates every year, against America's 50,000. But the quality is mixed. The best Indian IT firms fight over the top 30,000–40,000 graduates, a pool in which foreign companies such as IBM and Accenture also fish. Wage inflation at Wipro and Infosys is running at 15–17% a year, and is likely to worsen. Assuming a supply of 40,000 decent IT engineers a year, McKinsey's Diana Farrell thinks that India will "not even come close" to meeting the demand for 1m offshore IT and software workers her company forecasts for 2008.

The supply of top-quality Indian MBAs is also thinner than it might look at first sight. Indian business schools produce about 90,000 graduates a year, but everybody fights over the top 5,000 from the six state-run Indian Institutes of Management. "I'm afraid that for some of the private business schools it is two classrooms, 25 desktops, four faculty members, 600 books and you're away," sniffs one state-sector professor.

The biggest supply may be of BPO workers who do not need to use the telephone much: claims processors, credit-card administrators, health-insurance workers and so on. Indian universities churn out 2.5m graduates a year. Perhaps a quarter to half of these have the right skills to do this sort of BPO work, says NASSCOM's president, Kiran Karnik. To improve that ratio, he is working with India's University Grants Commission to have three-year degree courses supplemented by one-year technical certificates in IT or American accounting standards.

Mr Karnik thinks that the market itself will exact higher standards. The inferior private technical institutes and management schools that have sprung up since the government deregulated higher education in the 1990s charge about three times the fees of the elite state institutions, he says. No doubt the private schools will try to do better, but it will take time. Meanwhile, growing demand for offshore IT and call-centre workers is directing companies to other parts of the world.

Where to look next

The call-centre business in the Philippines is booming. China is attracting a healthy share of manufacturing-related R&D work: GE, Siemens and Nokia all do research there. Although China's IT industry is patchy and much less well organised than India's, this is likely to change in the next few years: China already churns out more IT engineers than India. Atos Origin, a big European IT-services firm, says it is more interested in China than in India because there is less competition for engineers.

The IT industry in eastern Europe and Russia is also scattered and poorly organised, but the talent is there if you look for it, says Arkadiy Dobkin. He is the head of Epam, an IT firm that claims to be the largest provider of offshore IT services in that part of the world, with over 1,000 engineers in Budapest, St Petersburg, Minsk and Moscow. "The engineers that Russia produces are comparable to India's," says Mr Dobkin. "The educational machine is still working." He reckons that a Russian or Hungarian IT engineer costs "about the same, or a little bit more" than an Indian engineer. American multinationals are already scouring the region for talent.

For the moment, India accounts for about 80% of the low-cost offshore market, and is probably exerting a stronger pull than ever. In the long run, however, it is sure to face hotter competition, especially from China and Russia. When it does, the abysmal quality of its infrastructure will become crucial. The most important thing to improve is India's airports, says Mr Murthy of Infosys: "The moment of truth comes when foreigners land in India. They need to feel comfortable." After airports, Mr Murthy lists better hotels, roads, schools and power supply.

Infosys's headquarters in Bangalore sit on 70 acres of pristine lawns and paths. The facilities include open-air restaurants, an amphitheatre, basketball courts, a swimming pool and even a one-hole golf course. "When we created this campus, we wanted everything to work as well as it does in America, to be as clean as America is," says Mr Murthy. But outside the perimeter walls, the place remains unmistakably India.

Faster, cheaper, better

India's emerging IT firms are trying to beat their western rivals on their home turf

CAN INDIA'S IT INDUSTRY do to the West's IT giants what Wal-Mart has done to rival retail firms, or Dell to computer-makers? The Indians talk a good game. "The productivity growth of Indian IT services is the highest in the world," says Mr Narayanan at Cognizant. He should know: one-third of his firm's employees are in America and two-thirds in India. Nandan Nilekani, the chief executive of Infosys, goes further. "Almost everything that is done can be done by us faster, cheaper and better," he says.

The argument for an Indian takeover of the world goes something like this. Like Dell and Wal-Mart, companies such as Infosys, TCS, Wipro and Cognizant source their offerings from poor, cheap countries. Wal-Mart has grown by adding Chinese-made toys, clothing and household appliances. Dell has added printers, hand-held devices and televisions to its line of made-in-Asia computers. In the same way, predict the Indian firms breezily, they will grow by adding new lines of IT services, offering global standards or better but produced at Indian costs. Investors understand this, say the Indians. Accenture's revenue is 14 times that of Infosys, but the American firm's market value is only one-third higher than that of its Indian competitor.

IBM and Accenture have been recruiting in India to lower their costs in areas where the Indian firms have grown fastest, such as maintaining popular business-software packages. But these global firms are so large (IBM employs 340,000 people; Accenture 100,000) that hiring even 10,000 extra staff in India has made little difference to their overall costs, most of which are still incurred in rich, expensive economies, the Indian firms point out gleefully. "The multinationals will never be able to restructure their costs fast enough to shift their centres of gravity," says Arindam Bhattacharya of the Boston Consulting Group in New Delhi.

Moreover, because the Indian firms know India better than their American and European rivals do, they can grow (and are indeed growing) more quickly and more cheaply in India than anyone else. This will lower their costs even further. "We're adding close to 5,000 people in

India this year," says Mr Narayanan. "No American company can do that." However, Accenture may have grown far more quickly in India than it can easily manage – though it bristles at the suggestion that it is finding India unusually difficult.

Wal-Mart sells commodities, such as microwave ovens at $28. In commodity businesses, the firm with the lowest price, which is often achieved by selling at the highest volume, wins the most customers. But not everything the IT industry sells is a commodity.

Layer cake

Broadly, the industry has three layers. The bottom one consists of businesses that have clearly become commodities. These are ruled by common standards, as in IT hardware manufacturing (where high-volume, low-cost Dell operates). A lot of this has moved to Asia.

The top layer is made up of tailored, bespoke technology services. Accenture, for instance, advertises work it has done for a large Australian casino to introduce a tracking technology, called Radio Frequency Identification, to improve the way the casino handles the 80,000 bits of staff clothing it has dry-cleaned every year. IBM is working with an American limousine-fleet company to introduce the same mathematical models the airline industry uses to route aircraft. Atos Origin, a European IT-services firm, is working with a British government agency, the Vehicle and Operator Services Agency, to equip its inspectors with hand-held computers to help them decide which passing vehicles to check. Because these services are tailored to meet the needs of individual customers, they are likely to continue to be provided close to the IT industry's biggest customers in America, Europe and Japan.

That leaves a large block of services sandwiched in the middle. These services are on their way to becoming commodities as shared standards spread. The ready adoption of a small number of business-software packages sold by firms such as SAP and PeopleSoft, for instance, is making the maintenance and even the installation of such software increasingly routine as these popular packages are becoming de facto standards. It is this large middle layer of services that is currently feeding the rapid growth of Indian firms such as TCS and Infosys.

Champions of the Indian firms look at the industry's employees and see a large bulge of people offering this middle layer of IT services, with a thinner sliver of business consultants doing the bespoke work on top. This makes them think that it should be far easier for the Indian firms to

Tomorrow's giants? **4.5**

India's top IT-services firms, 2004*

	Total revenue $bn	Market capitalisation Oct 29th, $bn	Employees '000
TCS	1.62	12.17	40.9
Wipro	1.34	10.13	32.0
Infosys	1.08	11.26	32.9
HCL Tech†	0.57	2.38	16.4
Satyam	0.57	2.62	15.6

*Financial years ending March †Financial years ending June
Sources: Company annual reports; Thomson Datastream

move up to that top layer by hiring consultants in America and Europe than for western IT firms to shift most of their employees (and their costs) from rich countries to poor ones. "About 20% of our value is added near our customers in America and Europe and 80% here in India," says Infosys's Mr Murthy. "If IBM wants to replicate this, it needs 80% of its employment in less developed countries as well."

This analysis neglects several important points. Perhaps the most crucial of these is that patterns of demand in the IT industry have shifted in the past, and may well do so again. Ten years ago customers spent a much bigger chunk of their IT budgets on computer hardware than they do now. Between 1993 and 2001, calculates Catherine Mann of the Institute for International Economics, spending on software and services grew by 12.5% a year, nearly twice as fast as hardware spending, pushing the share of software and services in overall expenditure from 58% to 69%.

As Ms Mann points out, the movement of IT hardware manufacturing to low-cost Asia helped to finance this shift in demand, because falling hardware prices freed up money to spend on software and services. Likewise, thinks Ms Mann, the migration of commodity IT services to low-cost places such as India will leave companies with more money to spend on the top-end bespoke services, which will help to expand this category of work.

If the world's IT giants want to remain big, they will have to change to meet changing demand. IBM has already performed this trick once. At the beginning of the 1990s, the company was mainly a hardware manufacturer. By the end of that decade, it had shifted much of its weight into IT services. Now, says IBM's Mr Harreld, the firm needs to move its high-cost employees into tailored services as commodity services migrate offshore.

The end of the beginning

Mr Harreld predicts that demand for such bespoke services will grow strongly, and that it will be many years before everything the IT industry sells becomes a commodity. To support his argument, he turns to Carlota Perez, an economic historian. In her book, *Technological Revolutions and Financial Capital* (Edward Elgar, 2002), Ms Perez traces five boom-and-bust cycles of technological innovation: the industrial revolution; steam and railways; steel, electricity and heavy engineering; oil, cars and mass production; and information technology and telecommunications.

In each age, argues Ms Perez, a phase of innovation, fuelled by hot money, has been followed by a financial bust, and then by an extended period in which the technology is deployed properly. Having just emerged from its bust, the information age is only at the beginning of this long deployment period, says Mr Harreld. Proper deployment, he argues, will require a large number of people working close to the industry's customers, in the way that IBM is doing for its limousine-fleet customer, or that Atos Origin is doing for Britain's vehicle-safety agency.

Two questions remain. The first is how long it will take for the large middle layer of services to become a commodity. If this happens too quickly, companies such as IBM, EDS and Accenture may find themselves overwhelmed by the pace of change, just as IBM nearly found itself ruined by the shift of IT manufacturing overseas in the early 1990s.

Of the three giants, EDS is in the weakest position. Having struggled with financial troubles and management turmoil at home, it has done little so far to counter the threat from Indian competitors, who are eating into large chunks of its business. Other smaller IT-services companies, such as BearingPoint and Capgemini, may also struggle with the shift of services abroad.

Most services in the middle layer, however, are likely to move offshore at a fairly manageable speed. That is because the IT organisations of most large companies tend to be a tangled mess of overlapping systems which go wrong so often that, as a practical matter, it will be hard to move IT work anywhere without fixing the systems first. To illustrate this point, Mr Harreld produces a diagram showing the different systems of one of IBM's customers, along with their interconnections. It is so intricate that it might pass for the design of a semiconductor chip. IBM itself runs 17,000 software applications, a figure that Mr Harreld thinks can comfortably shrink to 10,000 in due course.

The other big question is how easily companies such as Wipro, TCS and Infosys can expand into that upper crust of bespoke services that Mr Harreld predicts will flourish close to the industry's customers in rich countries. The Indian firms have lots of cash to spend: the cost of an

Indian programmer is so much lower than an American one that Wipro and Infosys are earning fat profits on lines of business that may be only just profitable for big western companies. So far, the Indians have spent their money cautiously, making small acquisitions and hiring the odd western consultant from rival firms.

If they are serious about taking on companies such as IBM and Accenture, the Indian firms will have to act more boldly. Yet buying or building people businesses of this kind is notoriously difficult. Time and again, and in all sorts of industries, from banking to telecommunications, America's and Europe's best managers have tried and failed miserably. Moreover, the competition is well entrenched. IBM, for example, has built up good relations with its customers over decades. The Indian companies may yet find that the only thing they can do faster and better on their rivals' home turf is to lose their shirts.

Into the unknown

Where will the jobs of the future come from?

"**H**AS THE MACHINE in its last furious manifestation begun to eliminate workers faster than new tasks can be found for them?" wonders Stuart Chase, an American writer. "Mechanical devices are already ousting skilled clerical workers and replacing them with operators ... Opportunity in the white-collar services is being steadily undermined." The anxiety sounds thoroughly contemporary. But Mr Chase's publisher, MacMillan, "set up and electrotyped" his book, *Men and Machines*, in 1929.

The worry about "exporting" jobs that currently grips America, Germany and Japan is essentially the same as Mr Chase's worry about mechanisation 75 years ago. When companies move manufacturing plants from Japan to China, or call-centre workers from America to India, they are changing the way they produce things. This change in production technology has the same effect as automation: some workers in America, Germany and Japan lose their jobs as machines or foreign workers take over. This fans fears of rising unemployment.

What the worriers always forget is that the same changes in production technology that destroy jobs also create new ones. Because machines and foreign workers can perform the same work more cheaply, the cost of production falls. That means higher profits and lower prices, lifting demand for new goods and services. Entrepreneurs set up new businesses to meet demand for these new necessities of life, creating new jobs.

As Alan Greenspan, chairman of America's Federal Reserve Bank, has pointed out, there is always likely to be anxiety about the jobs of the future, because in the long run most of them will involve producing goods and services that have not yet been invented. William Nordhaus, an economist at Yale University, has calculated that under 30% of the goods and services consumed at the end of the 20th century were variants of the goods and services produced 100 years earlier. "We travel in vehicles that were not yet invented that are powered by fuels not yet produced, communicate through devices not yet manufactured, enjoy cool air on the hottest days, are entertained by electronic wizardry that was not dreamed of and receive medical treatments that were unheard

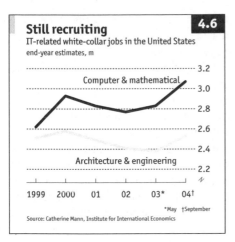

Still recruiting `4.6`
IT-related white-collar jobs in the United States
end-year estimates, m

Computer & mathematical

Architecture & engineering

1999 2000 01 02 03* 04†

*May †September
Source: Catherine Mann, Institute for International Economics

of," writes Mr Nordhaus. What hardy late 19th-century American pioneer would have guessed that, barely more than a century later, his country would find employment for (by the government's latest count) 139,000 psychologists, 104,000 floral designers and 51,000 manicurists and pedicurists?

Even relatively short-term labour-market predictions can be hazardous. In 1988, government experts at the Bureau of Labour Statistics confidently predicted strong demand in America over the next 12 years for, among others, travel agents and petrol-station attendants. But by 2000, the number of travel agents had fallen by 6% because more travellers booked online, and the number of pump attendants was down to little more than half because drivers were filling up their cars themselves. Of the 20 occupations that the government predicted would suffer the most job losses between 1988 and 2000, half actually gained jobs. Travel agents have now joined the government's list of endangered occupations for 2012. Maybe they are due for a modest revival.

You never know

The bureau's statisticians are now forecasting a large rise in the number of nurses, teachers, salespeople, "combined food preparation and serving workers, including fast food" (a fancy way of saying burger flippers), waiters, truck drivers and security guards between 2004 and 2012. If that list fails to strike a chord with recent Stanford graduates, the bureau also expects America to create an extra 179,000 software-engineering jobs and 185,000 more places for computer-systems analysts over the same period.

Has the bureau forgotten about Bangalore? Probably not. Catherine Mann of the Institute for International Economics points out that the widely quoted number of half a million for IT jobs "lost" to India in the past few years takes as its starting point the year 2001, the top of the industry's cycle. Most of the subsequent job losses were due to the recession in the industry rather than to an exodus to India. Measured from 1999 to 2003, the number of IT-related white-collar jobs in America has risen (see Chart 4.6 on the previous page).

Ms Mann thinks that demand will continue to grow as falling prices help to spread IT more widely through the economy, and as American companies demand more tailored software and services. Azim Premji, the boss of Wipro, is currently trying to expand his business in America. "IT professionals are in short supply in America," says Mr Premji. "Within the next few months, we will have a labour shortage."

If that seems surprising, it illustrates a larger confusion about jobs and work. Those who worry about the migration of white-collar work abroad like to talk about "lost jobs" or "jobs at risk". Ashok Bardhan, an economist at the University of California at Berkeley, thinks that 14m Americans, a whopping 11% of the workforce, are in jobs "at risk to outsourcing". The list includes computer operators, computer professionals, paralegals and legal assistants. But what Mr Bardhan is really saying is that some of this work can now also be done elsewhere.

What effect this has on jobs and pay will depend on supply and demand in the labour market and on the opportunity, willingness and ability of workers to retrain. American computer professionals, for instance, have been finding recently that certain skills, such as maintaining standard business-software packages, are no longer in such demand in America, because there are plenty of Indian programmers willing to do this work more cheaply. On the other hand, IT firms in America face a shortage of skills in areas such as tailored business software and services. There is a limited supply of fresh IT graduates to recruit and train in America, so companies such as IBM and Accenture are having to retrain their employees in these sought-after skills.

Moreover, Mr Bardhan's list of 14m jobs at risk features many that face automation anyway, regardless of whether the work is first shipped abroad. Medical transcriptionists, data-entry clerks and a large category of 8.6m miscellaneous "office support" workers may face the chop as companies find new ways of mechanising paperwork and capturing information.

Indeed, the definition of the sort of work that Indian outsourcing firms are good at doing remotely – repetitive and bound tightly by rules – sounds just like the sort of work that could also be delegated to machines. If offshoring is to be blamed for this "lost" work, then mechanical diggers should be blamed for usurping the work of men with shovels. In reality, shedding such lower-value tasks enables economies to redeploy the workers concerned to jobs that create more value.

Stuart Chase understood the virtuous economics of technological change, but he still could not stop himself from fretting. "An uneasy suspicion has gathered that the saturation point has at last been reached," he reflected darkly. Could it be that, with the invention of the automobile, central heating, the phonograph and the electric refrigerator, entrepreneurs had at long last emptied the reservoir of human desires? He need not have worried. Today's list of human desires includes instant messaging, online role-playing games and internet dating services, all unknown in the 1920s. And there will be many more tomorrow.

Sink or Schwinn

Sourcing from low-cost countries works only in open and flexible labour markets. Europe's are neither

WHEN HAL SIRKIN was growing up in 1960s America, the bicycle that every regular American child wanted was a Schwinn. In 1993, Schwinn filed for bankruptcy. The firm had been overtaken by imported Chinese bicycles. In 2001, a company called Pacific Cycle bought the Schwinn brand out of bankruptcy. Pacific Cycle, now owned by a Canadian consumer-goods firm called Dorel Industries, says the secret of its success is "combining its powerful brand portfolio with low-cost Far East sourcing". Schwinn bicycles now line the aisles at Wal-Mart.

Mr Sirkin is a consultant with the Boston Consulting Group who helps his customers do what Pacific Cycle has done to Schwinn: move production to East Asia, especially to China. Wal-Mart buys $15 billion-worth of Chinese-made goods every year. Obtaining goods and services from low-cost countries helps to build strong, growing companies, such as Dorel Industries, and healthy economies. But the Schwinn story also contains the opposite lesson: failing to buy in this way can seriously damage a company's health.

Sourcing from low-cost countries brings many economic benefits. Cheaper labour brings down production costs. This keeps companies competitive, raises profits and reduces prices as firms pass their lower costs on to their customers. Higher profits and lower prices lift demand and keep inflation in check. Companies spend their profits on improving existing products or introducing new ones. Customers buy more of the things they already consume, or spend the money on new goods and services. This stimulates innovation and creates new jobs to replace those that have gone abroad.

Moving work abroad may also help to speed up innovation directly, as American, European and Japanese companies get some of their R&D done by Chinese, Russian or Indian engineers. Randy Battat, the boss of Airvana, a telecoms-equipment start-up, has spent the past 18 months setting up an R&D centre for his company in Bangalore. This will complement the work of Mr Battat's engineers in Chelmsford, Massachusetts. The ones working in America will develop the next generation of the company's technology. The Bangalore centre will elaborate Airvana's existing technology. "They are adding bells and whistles that could not be added otherwise because it would not be cost-effective," says Mr Battat.

By making IT more afford-able, sourcing from cheaper countries also spreads the pro-ductivity-enhancing effects of such technology more widely through the economy. Ms Mann of the Institute for International Economics cal-culates that globalised pro-duction and international trade have made IT hardware 10–30% cheaper than it would

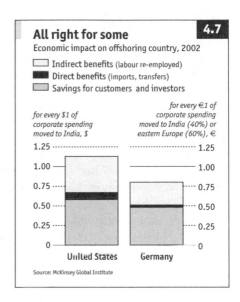

All right for some   `4.7`

Economic impact on offshoring country, 2002

☐ Indirect benefits (labour re-employed)
■ Direct benefits (imports, transfers)
☐ Savings for customers and investors

Source: McKinsey Global Institute

otherwise have been. She reckons that this price reduction created a cumulative $230 billion-worth of additional GDP in America between 1995 and 2002 as more widespread adoption of IT raised productivity growth. Sourcing IT services (which account for 70% of overall corporate spending on IT) from countries such as India will create a "second wave of productivity growth", predicts Ms Mann, as cheaper IT spreads to parts of the economy that have so far bought less of it, such as the health-care industry and smaller companies.

McKinsey calculates that for every dollar American firms spend on service work from India, the American economy receives $1.14 in return. This calculation depends in large part on the ability of America's economy to create new jobs for displaced workers. America's labour market is a miracle of flexibility: it creates and destroys nearly 30m jobs a year.

However, in countries such as Germany, France and Japan a combination of social legislation, stronger trade unions, regulations and corporate-governance arrangements make employment practices more rigid and sometimes keep wages higher than they would otherwise be. This reduces demand for labour and pushes unemployment higher. According to McKinsey, in Germany, the re-employment rate for IT and service workers displaced by sourcing from low-cost countries may be only 40%. As unemployment at home rises, that process could actually make Germans poorer (see Chart 4.7).

Reluctant Europeans

Udo Jung of the Boston Consulting Group says that, by and large, Germans accept that manufacturing companies such as Hella, Bosch and Siemens must get supplies from China. Degussa, a chemicals manufacturer, recently invited its workers' council on a trip to China. The idea was to take emotion out of the debate, says Mr Jung. Nor do continental Europeans seem bothered about white-collar work being done in low-cost countries. But that may be because they are doing so little of it.

At present, perhaps 80–90% of the service work being done remotely in India comes from either America or Britain, with which the country has linguistic and cultural links. Such links are absent from its relationship with Germany or France. Germany, like America, introduced a special visa programme for Indian IT workers in the 1990s as its domestic supply of engineers ran dry. But most Indians that went to work in Germany failed to learn the language and came back again, says Infosys's Mr Murthy. The opposite is true of Indians in America. Those who have gone there to work or study are often reluctant to return home to their families.

Cultural ties appear to be important in forming business relationships in remote-service work, says Rajendra Bandri of the Indian Institute of Management in Bangalore. Mr Bandri has studied five examples of European firms outsourcing white-collar work to Sri Lanka. In each case, they chose that country because a well-placed Sri Lankan worked for the European firm, says Mr Bandri.

Eastern Europe and Russia, which brim over with skilled, underemployed engineers, present fewer cultural barriers for European companies. French is spoken in Russia, German in Hungary and elsewhere. Yet neither German nor French firms have yet shown much appetite for buying services work from their neighbours, either. Arkadiy Dobkin, the boss of Epam, which claims to be the largest supplier of IT services from eastern Europe and Russia, is based in Princeton, New Jersey, rather than in Paris or Berlin.

Beyond economics

A survey of 500 European firms in the summer of 2003 by IDC, a research firm, found that only 11% of its sample were sourcing IT work from low-cost countries, and that nearly 80% would not even consider doing so. Attitudes were hardest in Italy, where 90% of firms were against the idea, followed by France and Germany. An American study released at the same time by Edward Perlin Associates, a consulting firm, found that around 60% of the companies it surveyed had some of their IT work done in low-cost countries.

In continental Europe, companies may outsource for reasons that have little to do with favourable economics, says Francis Delacourt, the head of outsourcing at Atos Origin. In what he describes as "social outsourcing", firms such as Atos Origin may take on surplus IT employees from companies that no longer need them. Europe-wide social legislation requires the new employer to provide the same wages and benefits as the old one. The alternative is costly redundancy. Mr Delacourt says this works for his company, up to a point, because demand for IT workers in Europe is growing, and Atos Origin has found ways to re-employ such people profitably. But he concedes that his company needs to be careful not to take on too many.

How well this system stands up to competition from India is anybody's guess. A manager at one firm in Europe privately muses that Germany, France and other countries might introduce barriers to IT imports to counter the threat to their domestic employment. If McKinsey is right and sourcing from abroad does make unemployment in Germany and elsewhere worse, protectionist sentiment will grow.

In the end, Europe's big service firms are likely to get round to sourcing production from abroad, as its manufacturing companies have already done. But by that time, says Andrew Parker of Forrester, British and American companies will already have developed much stronger ties with India and other cheap countries, and costs will have risen. This will especially hurt Europe's big financial firms: the biggest banks spend billions of dollars a year on IT. Mr Parker speculates that some European financial firms could be so badly damaged by this loss of competitiveness that they may fall into the arms of fitter American and British rivals. Schwinn could tell them all about it.

A world of opportunity

Why the protectionists are wrong

IN 2004, A GROUP OF POLITICIANS from Britain's left-of-centre Labour Party made a field trip to EXL Service, an Indian outsourcing firm in Delhi. Its charming boss, Vikram Talwar, must have worked wonders. On their return, the politicians chided Britain's trade unions for being negative about sourcing work from poor countries, and praised EXL Service's facilities for its workers. These included a health clinic, a gym and a good staff canteen. Laura Moffat, one of the politicians, approvingly told the *Financial Times*: "The benefits EXL offered its employees would be a wish-list for us in Britain."

More often than not in the past few years, public champions of outsourcing have found themselves bullied into silence. The chairman of President George Bush's Council of Economic Advisers, Gregory Mankiw, got howled off the stage in 2004 when he dared to defend the practice. Lou Dobbs, a TV news anchorman who names and shames unpatriotic American firms that hire workers abroad, is hawking around a new book, *Exporting America: Why Corporate Greed is Shipping American Jobs Overseas* (Warner Business Books, 2004).

Such attacks have instilled caution in some of the big technology firms: IBM, for instance, no longer likes to talk publicly about the growth of its business in India. Yet the backlash against outsourcing has been less violent than people like Mr Dobbs might have hoped; indeed, as the reaction of Mr Talwar's British visitors show, outsourcing is beginning to win support in unexpected quarters.

Protectionists are finding it hard to argue that "corporate greed" is draining jobs from Britain and America when those two economies are close to full employment. More awkwardly still, the very industries said to be badly hurt by the migration of jobs overseas report a shortage of workers at home. Most of the jobs created in India are either in call-centres or at IT firms. But call-centre companies in both Britain and America suffer from rising staff turnover and struggle to recruit more people. Britain's Call-Centre Association, a trade lobby, thinks that employment in the industry in Britain will rise in the next few years; in the United States, call-centre employment is expected to decline slightly.

As IT spending recovers from recession, labour markets in America and Europe are becoming tighter in this industry too. Not many students in rich countries choose to study engineering at college. Even a modest rise in the demand for IT workers in rich countries will create shortages – and therefore openings for Indian, Chinese and Russian engineers.

In the longer run, ageing populations in rich countries will mean labour shortages in many industries. Sourcing some of the work from abroad will ease the problem. It will also help to lift productivity among rich-country workers who will have to support larger numbers of older people. Moreover, it could help to lower some of the costs of ageing populations, especially in health care. America's health-care spending is rising at 12% a year, far faster than GDP. Farming out the huge job of administering this system to lower-cost countries would restrain such spending. Trade has the same sort of effect, and Americans think nothing of shopping online for cheaper drugs from Canadian pharmacies. Yet, as McKinsey's Diana Farrell points out, it is precisely the supporters of drug imports (and haters of big business) who complain most about jobs going to India.

Anti-globalisers claim that multinational firms that obtain goods and services in low-cost countries exploit the poor by putting them to work in sweatshops. Trade unions and industrial lobbies use such arguments to make their demands for protection look less self-interested, and guilt-wracked American and European *bien pensants* swallow them whole.

The spread of global sourcing may help to unpick these politics. The smartly dressed, brand-conscious young men and women who stroll around the lush technology parks of Bangalore are patently not some new underclass. New wealth in the East will help to expose old protectionist politics in the West. That might provide globalisation with a new legitimacy and moral strength.

Although the opportunity to source large amounts of white-collar work from low-cost countries has arisen quite suddenly, the work will in fact move over gradually. This will give rich economies time to adjust to new patterns of work, and should keep the politics of change manageable. But from time to time, ugly protectionism is sure to flare up again.

Take it gently

A sudden increase in global competition could force faster and deeper restructuring in rich countries. Big IT-services firms such as IBM and Accenture have scrambled to hire tens of thousands of new employees in India to compete with Indian IT firms such as Wipro and TCS. This could happen in other industries, too, as India becomes expert at providing outsourced banking, insurance and business services.

Office workers everywhere are likely to be discomfited by the rise of Indian firms that promise to do white-collar work cheaper, faster and better. Just as the Japanese carmakers licked Detroit into shape, India is going to change life on the cubicle farm forever. So far only American and British firms have sourced much work from low-cost countries, but other rich economies such as France, Germany, Italy and Japan will eventually have to follow as British and American firms reduce their costs and make their rivals look vulnerable. In Japan, France and Germany, this could lift high levels of unemployment (disguised in Japan; explicit in France and Germany) higher still if rigid, unreformed labour markets continue to deny displaced workers new jobs. This is likely to fuel protectionism and cause a backlash.

That may be all the more reason to reassert both the economic and the moral case for free trade. Buying goods and services from poor countries is not only hugely beneficial to rich countries' economies, it can also provide opportunities for millions of people in poor countries to lift themselves up and improve their lives. It is a game in which everybody can win.

The material on pages 112–46 first appeared in a survey in *The Economist* in November 2004.

PART 2

THE SHIFT TOWARDS CONSUMER ELECTRONICS

Part 2 consists of three collections of articles about different fields of consumer electronics. The first section, about mobile phones, looks at the emergence of the mobile handset as a truly personal computer and the battles that have followed as different firms compete to provide handsets and related services, most notably through the introduction of third-generation (3G) networks. The enormous variety of mobile-phone designs is also considered, along with their social impact, both good and bad. The second section covers video games, which are gradually establishing themselves as a branch of the mainstream entertainment industry, alongside movies and music. Games are also being taken increasingly seriously as training tools, and not just by airline pilots, as the growing power of games consoles enables them to create increasingly photo-realistic images. The third section looks at the migration of technologies from the computer industry – such as flat-panel displays and wireless networks – into the domestic sphere, and the rise of the "digital home". This could eventually result in entertainment info-nirvana, but only if content providers and technology firms can agree on appropriate standards.

5
MOBILE PHONES

Computing's new shape

As two industries collide, a new kind of computer may emerge

"A COMPUTER ON EVERY DESK and in every home." This was Microsoft's mission statement for many years, and it once sounded visionary and daring. But today it seems lacking in ambition. What about a computer in every pocket? Sure enough, Microsoft has duly amended its statement: its goal is now to "empower people through great software, anytime, any place on any device". Being chained to your desktop is out: mobility is in. The titan of the computer industry has set its sights on an entirely new market.

It is not alone. Dell, Hewlett-Packard and other computer-makers have diversified into handheld computers, which increasingly have wireless connectivity and even full-fledged mobile telephony built in. The Palm Pilot, originally an electronic organiser, has metamorphosed into the Treo, a far more elaborate device which also incorporates a mobile phone, e-mail and wireless internet access.

As the computer industry tries to cram PCs into pocket-sized devices, the mobile-phone industry has arrived at the same point – but from the opposite direction. Most mobile phones now have colour screens and internet access. Some, aimed at business users, have tiny keyboards to facilitate the writing of e-mails; others are specifically designed for music playback or gaming. Today's phones have as much computing power as a desktop computer did ten years ago.

In short, the once-separate worlds of computing and mobile telephony are now colliding, and the giants of each industry – Microsoft and Nokia, respectively – are engaged in a fight for pre-eminence (see pages 154-61). Both camps are betting that some kind of pocket communicator, or "smartphone", will be the next big thing after the PC, which has dominated the technology industry ever since it overthrew the mainframe some 20 years ago. Admittedly, the two camps have different ideas about how such devices should be built. The computer industry believes in squeezing a general-purpose computer into a small casing; the mobile-phone industry takes a more gentle, gradualist approach of adding new features as consumers get used to existing ones. But are the two sides right about the future of computing in the first place?

The answer is probably yes, even though it is too early to be absolutely sure. As they search for new growth, both industries are certainly acting on that assumption. In the case of computers, sales of PCs have levelled off and corporate spending has slowed, so bets are being placed on mobile personal devices. For mobile phones, revenues from voice calls are now flat, so new data services such as photo-messaging, gaming and location-based information are looked to as the most promising source of growth.

Inevitably, there have been mis-steps already; the most obvious was the fiasco of European operators' attempts to launch "third-generation" (3G) mobile networks. The operators' willingness to pay vast amounts of money for licences to operate 3G networks shows how fervently they believed that the convergence of computers and phones was the next big thing. Even so, they paid too much: over €100 billion ($125 billion) in all. 3G networks started to be switched on in earnest only in 2004, after repeated delays, and it is still unclear whether they were a good investment (see pages 162–9). Similarly, handheld computers, also known as personal digital assistants (PDAs), appear to have limited appeal; annual sales are flat at around 10m units.

Yet the trend remains clear. Mainframes ruled the computer industry until the rise of the PC; another 20 years on, the PC's reign now seems to be coming to an end. Previous generations of computers live on – mainframes are widespread, and PCs are certainly not going away – but each successive generation of computing devices is smaller, more personal and more numerous than its predecessor. Mainframes filled whole rooms, and belonged to single companies. PCs sit on desks and are used by individuals or households. Phones are truly personal pocket-sized devices that are carried everywhere. More than a billion people around the world now have one.

The switch to mobile devices is thus a logical long-term step. Moreover, the earliest incarnations of a technology, with all its snafus, are not always an accurate guide to its subsequent development. The short-term impact of a new technology is usually overstated; the long-term benefit is often underestimated. Consider the earliest PCs, 20 years ago. They were hardly consumer products, yet they evolved into something with far broader appeal.

Today's smartphones and handheld computers are at a similar stage of development. Their makers do not claim to have all the answers, and are hedging their bets. The chances are that a variety of devices will emerge, each appealing to a different type of user. Microsoft is pursuing both smartphones and slate-like handheld computers. Nokia has split its handset division into several "mini-Nokias", each concentrating on a different market segment, while sharing research, development and manufacturing facilities. And entirely new devices have appeared from companies such as Research in Motion, the Canadian firm behind the BlackBerry e-mail device, and Danger, a Silicon Valley firm with a pocket communicator that is neither a jazzed-up phone nor a scaled-down PC, but a genuine hybrid of the two.

Looking for the next Microsoft

If this is the next stage in the evolution of computing, one obvious question arises: which firm will dominate it, as IBM dominated the mainframe age, and Microsoft the PC era? The answer is that there is unlikely to be a single winner this time around. IBM ruled in mainframes because it owned the dominant hardware and software standards. In the PC era, hardware became an open standard (in the form of the IBM-compatible PC), and Microsoft held sway by virtue of its ownership of Windows, the dominant software standard. But the direction of both computing and communications, on the internet and in mobile telecoms, is towards open standards: communication devices are less useful if they cannot all talk to each other. Makers of pocket communicators, smartphones and whatever else emerges will thus have to compete on design and branding, logistics and their ability to innovate around such open standards.

These considerations seem to favour Nokia more than any other company. But Nokia faces a direct challenge as Microsoft leads the computer industry on to its turf; its continued dominance of the mobile-phone industry is by no means assured, since it is not based on the ownership of proprietary standards. Microsoft, for its part, will try to exploit its dominance of the PC industry to help force its way into the new market. But it may well fail. Either way, there will be no need this time round for any repeat of the long-drawn-out antitrust cases, against first IBM and then Microsoft.

Instead, the collision of the computing and mobile-phone industries seems likely to lead to a surge of innovation, as the two camps fight it out to create a truly personal computing and communications device, with far wider appeal than the misleadingly named personal computer. And as these titans slug it out, it will be consumers who emerge as the winners.

POSTSCRIPT

Since this article appeared in 2002, smartphones have become increasingly elaborate and have continued to incorporate features (such as hard disks and Wi-Fi connectivity) from PCs. Around 26m smartphones were sold in 2004, a mere 4% of all mobile phones sold, but sales are growing at around 30% a year, according to Jupiter Research, a consultancy. Nokia remains the leading maker of smartphones, with a market share of 50% in 2004, far ahead of its rivals palmOne (maker of the Treo) and RIM (maker of the BlackBerry). Microsoft has yet to make a significant impact in this market.

The material on pages 150–3 first appeared in *The Economist* in November 2002.

Battling for the palm of your hand

Just as mobile phones have changed dramatically in recent years, the industry that makes them is being transformed too

NEXT TIME YOU PICK UP YOUR MOBILE PHONE, try to imagine how futuristic it would look to someone from the mid-1990s. Back then, mobile phones were far less sophisticated devices. Brick-like, they had tiny monochrome screens and ungainly protruding aerials, and they were only used for one thing: talking to other people. Today's latest models, in contrast, are elegantly shaped pocket computers. Your current handset may well have a large colour screen and a built-in camera; as well as being a telephone, it can send and receive text messages, and may also serve as an alarm clock, calendar, games console, music player or FM radio.

The mobile phone has become a uniquely personal item: many people take theirs with them even when leaving wallets or keys behind. Some phones designed for business users can send and receive e-mail, and have tiny keyboards; others aimed at outdoor types have built-in torches; still others have satellite-positioning functions, high-resolution cameras with flash and zoom, and even the ability to record and play video clips. Clearly, phones ain't what they used to be.

This spectacular outward transformation of the mobile phone is being reflected by an internal transformation of the industry that makes what have now become the most ubiquitous digital devices on the planet. Over half a billion mobile phones are sold every year, and despite sluggishness in other parts of the technology industry, the number continues to grow (see Chart 5.1). Sales are being driven, in part, by the surge of new subscribers in the developing world, particularly in India and China. In the developed world, meanwhile, where markets are so saturated that most adults already carry a mobile phone, existing subscribers are switching in droves to today's more advanced models. Globally, the number of mobile phones in use, at around 1.4 billion, overtook the number of fixed-line phones in 2003.

No wonder so many firms now want a piece of the action. The mobile phone sits at the intersection of three fast-moving industries: it is a communications device, computer and, with the addition of new media functions, consumer-electronics product. Indeed, it is the bestselling device in all three categories.

As a result, the firms that have historically dominated the industry – large, specialised firms such as Nokia and Motorola – now face a host of new challengers as well as opportunities. The desire for "ownership" of each mobile-phone subscriber poses another threat to the incumbent handset-makers, as mobile-network operators seek to promote their own brands and to differentiate themselves from their rivals. The result is a little-seen, but almighty, struggle for control of a $70 billion industry: a battle, in short, for the palm of your hand.

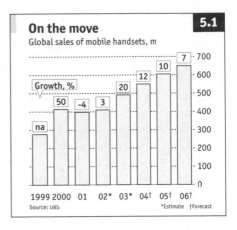

On the move 5.1
Global sales of mobile handsets, m
Growth, %
Source: UBS   *Estimate †Forecast

Making a mobile phone used to be so difficult that it was the exclusive province of a few specialist companies. It required expertise in an enormous range of areas, from the design of radio chips and software to the integration of electronic components and the styling of the case. Then, since the handsets had to be cranked out in large quantities, there were the problems of running an efficient manufacturing process and complex supply chain, as well as promoting the finished products to a mass consumer market. Furthermore, a company could not just make handsets: to be taken seriously by the mobile-network operators, and ensure everything worked properly, it also had to manufacture the much larger and more complex base-stations that are used to provide mobile-phone coverage.

All these requirements meant that the industry came to be dominated by large, vertically integrated firms such as Nokia, Motorola and Ericsson. "For many firms good at low-cost electronics, the barrier to entry was simply too high," says Tony Milbourn of TTPCom, a British firm that designs and licenses hardware and software components for mobile phones.

But the situation has changed. Radio chips can now be bought off the shelf, as can the software required to make a mobile phone work. Manufacturing can be outsourced to an "electronic-manufacturing services" (EMS) firm. Some of these have started to design as well as build handsets; these "original design manufacturers" (ODMS) sell their fin-ished phones to other firms, which in turn sell them under their own brands. Meanwhile, a flourishing ecosystem has sprung up of small firms specialising in areas such as handset design, chip design, testing and software. TTPCom, for example, provides the software that enables Sharp camera-phones and BlackBerry wireless e-mail devices to send and receive data over mobile-phone networks.

In other words, the barriers to entry have fallen. Hardware and soft-ware have, to some extent, been commoditised, and there is far more scope for outsourcing of design and manufacturing than there used to be. This has allowed ODMS, consumer-electronics firms and even start-ups to enter the handset business. "Anybody with the right financial backing can break into the phone business now," says Ben Wood, an analyst at Gartner, a consulting firm. The old vertical industry model has been undermined. And it is the rise of the ODMS in particular that is doing the most to disrupt the industry's established order.

Oh Dear Me?

Most ODMS – the biggest are BenQ, Arima and Compal – are based in Taiwan, though there are others in China and South Korea too. All of them design and build handsets for better-known firms, which simply apply their own branding to the finished phones and sell them as their own. The irony is that at the moment the ODMS' biggest customers are the incumbent handset-makers. Arima, for example, makes phones for Sony Ericsson (a handset joint-venture between Sony of Japan and Eric-sson of Sweden), while BenQ and Compal make several models for Motorola. Siemens, Toshiba and Panasonic also rely on ODMS to pro-duce some of their phones.

Using an ODM, at least to make some models, has several advantages for the established handset-makers, says Adam Pick of iSuppli, a market-research firm. It lets them fill gaps in their product lines quickly and cheaply; it saves money on research and development; and it means the ODM takes on some of the business risks associated with fluctuations in component supply and end-user demand. Northstream, a Swedish consultancy, predicted that the proportion of handsets pro-duced by ODMS would grow from 9% in 2002 to around 30% by 2005.

But the growing importance of ODMS also poses a long-term threat to the established handset-makers. Motorola, for example, spent a few months polishing an original design from Chi Mei, a Taiwanese ODM, to produce its MPX200 handset, the specifications for which then belonged to Motorola. This approach allows fast time-to-market and means the ODM cannot supply the improved design to rivals.

The risk, notes Mr Wood, is that this process also educates the ODM. By getting too cosy with ODMs, big handset-makers could end up eroding their current technological lead and unwittingly nurturing their own future competitors. Already, some ODMS – BenQ is the most prominent example – are selling handsets under their own brands in some countries. In March 2004 BenQ overtook Nokia to become the number two in the Taiwanese handset market.

That is not the only risk. The rise of the ODMs also allows operators to bypass the established handset-makers and produce their own "operator-specific" handsets. This lets them apply their own branding to the handsets and helps them to differentiate themselves from rival operators.

For example, Orange, a European mobile operator owned by France Telecom, sells own-brand smartphones that are built by HTC, yet another Taiwanese ODM, and powered by software from Microsoft. Figures from Orange suggest that these smartphones increase the average revenue per user (ARPU), a crucial industry yardstick, by around €15 ($18) a month. By closely tying a handset with services (in this case mobile e-mail and web browsing), operators can increase revenue and discourage subscribers from defecting to other operators – two key goals.

The success of the Orange handset also showed that operator-specific handsets could deliver results, says John Pollard, director of business strategy at Microsoft's mobile division. "They shipped the thing, and it didn't break," he says. This has emboldened other operators to follow suit.

Big handset-makers were initially reluctant to modify their handsets for individual operators, since this can reduce their economies of scale and dilute their own brands. But the threat of operator-specific handsets supplied by ODMs has now forced the established suppliers to become more flexible. Motorola, for example, produced a special version of its v500 handset specifically for Vodafone, the world's largest mobile operator, notes Mr Wood. Nokia's reluctance to be flexible – as the leading handset-maker, with the strongest brand, it has the strongest bargaining position with the operators – may have contributed to its unexpected stumble in April 2004. Its share price fell sharply after the company announced that sales had fallen in the first quarter of 2004.

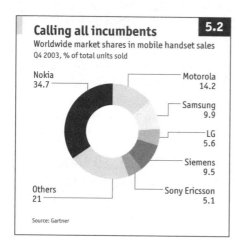

Calling all incumbents `5.2`

Worldwide market shares in mobile handset sales
Q4 2003, % of total units sold

Nokia 34.7
Motorola 14.2
Samsung 9.9
LG 5.6
Siemens 9.5
Others 21
Sony Ericsson 5.1

Source: Gartner

Here comes the gorilla

The rise of the ODMs also benefits Microsoft, which had been having great difficulty breaking into the mobile-phone business. Rather than use Microsoft's smartphone software, the established handset-makers set up Symbian, a consortium, to produce an alternative. (They were worried by what happened in the PC industry, where Microsoft established a software monopoly and reduced PC-makers to merely efficient box-shifters with almost identical products.)

But Microsoft got its foot in the door by teaming up with ODMs to produce operator-specific handsets, thus bypassing the established handset-makers. Since then, Motorola and South Korea's Samsung have licensed Microsoft's software for use on some of their phones. Does all this mean the handset industry could end up going the way of the PC industry after all?

It seems unlikely. For one thing, mobile phones are far more personal items than PCs; in effect, they have become fashion items. So far, there is no sign of a Microsoft or Intel-like monopoly in any of the new horizontal layers of the handset industry; the most important standards are open, and exist at the network layer. The power of the mobile-network operators has no parallel in the PC industry; internet-service providers have very little clout. Instead, a better analogy for the mobile-phone industry's new structure would seem to be carmaking.

Like handset-makers, carmakers used to be entirely vertically integrated. But now there is a complex mix of different approaches. Some carmakers outsource the manufacturing of particular components (engines, for example, or lighting systems); Motorola does the same for the lids of its flip-phones. "For big sub-assemblies, it's very similar to the automotive industry," says Tom Lynch, president of Motorola's mobile-phone business. "That's the way manufacturing is going."

Similarly, the design and manufacturing of some cars is outsourced altogether to ODM-like specialists. BMW, for example, got Magna Steyr, an Austrian firm, to design and build its X3 sports-utility vehicle, partly in order to reduce its time-to-market in a competitive sector. But BMW also makes cars in the old-fashioned, vertically integrated way.

Another parallel between the two industries is the use of "platforms" – underlying common designs that allow components to be shared between outwardly different products. This has the benefit of reducing costs, but it can be taken too far if it results in a bland product line-up. Nokia seems to have run into this problem, as Volkswagen did in the car business.

If the car industry is any guide, then, the likely outcome is that the handset business will shift from vertical integration to a series of horizontal layers: chips, software, manufacturing, design and branding. But unlike in the PC business, no single company will dominate any one layer. Specialists will proliferate; and many firms will choose to compete in more than one layer at once, depending on where their competitive advantage lies. On the spectrum between total vertical integration and total commoditisation, handset-makers hope to end up in the middle, says Vasa Babic of Mercer, a consultancy. They will then be able to reap the benefits of commoditisation, such as lower component prices, without ending up like the PC-makers.

Now what?

The shift away from the old vertical model is causing the incumbent handset-makers to change their strategies in varying degrees. Being able to design your own radio chips is now less important than it used to be; so is owning all of your own manufacturing capacity, or making all your own software. So the incumbents have shifted towards using off-the-shelf chips and software, and to increasing their use of outsourced manufacturing, in the form of both EMS and ODM firms.

Motorola and Sony Ericsson now outsource around 35% of their manufacturing, and neither firm any longer designs its own radio chips. Siemens has taken a similar approach, mixing in-house products with ODM handsets. Nokia, however, insists that its sheer size – its market share is around 35% – means it can still compete in every layer, from chip design to branding. The company relies less on outsourcing than rivals, using it to respond to variations in demand and to benchmark the efficiency of its own manufacturing. Even so, says Olli-Pekka Kallasvuo, head of Nokia's mobile-phones business group, the firm takes a "pragmatic" view and might rely more on outsourced manufacturing in future. Samsung, the world's third-biggest handset-maker, is also sticking with a traditional vertical model.

But will all this be enough to fend off their new competitors? One problem for Nokia and Motorola, says Mr Wood, is that, while they are committed to competing in every market worldwide, their smaller rivals are able to "cherry pick" particular markets and product niches. ODMs already have more than 20% market share in Taiwan, and are now targeting specific markets elsewhere, such as Italy and certain countries in central Europe. This enables them to form relationships with big operators, who in turn can put more pressure on the established handset-makers.

Sendo, a British start-up, is an interesting example. It launched its first handsets in June 2001, and now claims a 5% market share in Britain, and 9% in Portugal, as a result of its willingness to customise phones for particular operators. For Virgin Mobile, for example, Sendo made a phone with a special "v" button. The firm is not an ODM, but concentrates on software, design and customisation, and outsources everything else. It is a new kind of handset-maker that could not have existed just a few years ago.

The growing popularity of outsourced manufacturing is by no means an unstoppable trend, however. Sony Ericsson used to outsource all of its manufacturing, but found its suppliers were unable to meet demand. Katsumi Ihara, the company's president, says about one-third of the company's handsets are now made by Flextronics, an EMS firm based in Singapore. "We are looking for the best mixture," he says. Similarly, Motorola went too far down the outsourcing route and has since retreated to gain more control: around 35% of its manufacturing is now outsourced, says Mr Lynch. Northstream predicts that, in the long run, ODMs will account for around half of all handsets manufactured.

Another extreme outcome – the complete disintermediation of handset-makers by operators – also seems highly unlikely. Operators like to have someone to blame when handsets go wrong, and with handsets becoming ever more complex and more reliant on software they are unlikely to want to take on servicing and support. So while some operators will use some own-branded ODM handsets in particular niches, they will not want to do away with traditional handset-makers altogether. "No operator wants to bet its future on sourcing all its handsets from China," says David Dean, an analyst at Boston Consulting Group.

Instead, the most likely outcome is a compromise in which the established handset-makers' power is reduced by an unspoken alliance between the operators and the new handset-makers. Co-branding, where handsets feature the logos of both operator and handset-maker, looks likely to become the norm. It is already widespread in America and parts of Asia, and is becoming more popular in Europe.

The company with the most to lose is Nokia, which has become so powerful that operators and rival handset-makers are keen to take it down a peg or two. "For Nokia to stay on top of the game it will have to adjust," says Brian Modoff, an analyst at Deutsche Bank. Unlike Microsoft and Intel in the PC business, Nokia is not protected by ownership of proprietary standards. To maintain margins and stay ahead of the industry's ever-faster product cycle, says Mr Modoff, it will have to stop doing everything itself. "They will be more of a brand, a design shop, rather than building everything," he says.

Nokia is doing its best to diversify, notably into mobile gaming with its N-Gage handset. At the same time, as handset technology is progressively commoditised, a strong brand will be increasingly important, and nobody has a stronger brand than Nokia. Its troubles may turn out to be a blip. But given the seismic shifts now under way in the industry, observes Mr Dean, with Nokia's market share so large "there's only really one way to go."

The material on pages 154–61 first appeared in The Economist in May 2004.

Vision, meet reality

After years of delay, third-generation (3G) mobile-phone networks are finally being switched on. How will the reality compare with the original vision?

THE BIGGEST EVER GAMBLE on the introduction of a new technology; an attempt to maintain growth in a maturing industry; or an industrial-policy fiasco? The introduction of "third-generation" (3G) mobile-phone networks around the world is all these things and more. In 2000, at the height of the dotcom boom, mobile operators around the world, but mainly in Europe, paid a total of €109 billion (then $125 billion) for licences to build and operate 3G networks, which offer higher performance and more capacity than existing second-generation (2G) networks. In part, the mobile operators were victims of their own hype. A report that year from the International Telecommunication Union, the industry's standards body, gives a sense of the high hopes for 3G:

> *The device will function as a phone, a computer, a television, a pager, a videoconferencing centre, a newspaper, a diary and even a credit card ... it will support not only voice communications but also real-time video and full-scale multimedia. It will automatically search the internet for relevant news and information on pre-selected subjects, book your next holiday for you online and download a bedtime story for your child, complete with moving pictures. It will even be able to pay for goods when you shop via wireless electronic funds transfer. In short, the new mobile handset will become the single, indispensable "life tool", carried everywhere by everyone, just like a wallet or purse is today.*

Dotcom mania aside, the industry had concluded that 3G networks would make possible new services to provide growth as its core business, voice telephony, matured. As the proportion of people with mobile phones has increased – it now exceeds 85% in much of the rich world – the average revenue per user (ARPU), a key industry metric, has levelled off. This is because the most valuable subscribers were the first to buy mobile phones; later adopters make fewer calls and spend much less. With subscriber numbers reaching saturation, at least in the rich world, the industry began casting around for new sources of growth, and fancy services such as video and internet access seemed the most promising prospects. Hence the appeal of 3G.

Even so, forking out €109 billion for 3G licences – plus roughly the same again between 2001 and 2007 to build the actual networks, according to predictions from iSuppli, a market-research firm – was an enormous gamble, arguably the biggest in business history. But in many cases operators had no choice. Several European countries held auctions for their 3G licences in which operators bid huge sums: in Britain and Germany, for example, operators ended up paying around €8 billion for each 3G licence. Why? Because with their 2G networks filling up, and with no additional 2G capacity on offer from regulators, operators felt compelled to buy 3G licences to ensure scope for future growth. Andrew Cole of A.T. Kearney, a consultancy, remembers when a client who was taking part in the auction received the order to "win the licence no matter what". The €109 billion was, in effect, a tax on the right to continue to do business. Few firms were brave enough to refuse to pay up.

So the 3G adventure got off to a bad start in Europe by nearly bankrupting the industry. Since 2000 most operators have written down the value of their 3G licences. Some even handed the licences back to the governments from which they bought them, rather than commit themselves to building expensive new 3G networks within strict time limits. (Reselling the licences was forbidden.) The whole episode is now something the industry would rather forget. "The spectrum auction is a nightmare the operators don't want to remember," says Mr Cole. "I haven't heard it mentioned in a long time."

Ready, steady, flop!

The pioneering launch of 3G services at the end of 2001 in Japan and South Korea, the world's two most advanced mobile markets, did little to lighten the mood. In both countries, operators were using 3G technologies different from the w-CDMA standard (which is also known as UMTS) being adopted in Europe. An unproven technology, w-CDMA was plagued by teething troubles: base-stations and handsets from different vendors would not work together reliably, and early handsets were bulky and temperamental. Operators postponed the launch of 3G services from 2002 to 2003 and then to 2004, though a handful chose to launch sometimes shaky 3G services earlier.

Yet in 2004, at last, the 3G bandwagon started to roll. According to figures from Deutsche Bank, there were 16 commercial 3G networks worldwide at the beginning of the year, and there were expected to be around 60 by the end of the year (see Chart 5.3). Matti Alahuhta, head of strategy at Nokia, the world's largest handset-maker, says the second half of 2004 will be seen as "the starting point for the global acceleration of 3G". The early, brick-like W-CDMA handsets have given way to much smaller, sleeker models. In Japan and Korea, sales of 3G handsets are booming. Even in America, that wireless laggard, 3G services have been launched in several cities, and the country's largest operators have committed themselves to building 3G networks.

Having swung too far towards pessimism, the industry is now becoming cautiously optimistic about 3G, says Tony Thornley, the president of Qualcomm, the firm that pioneered the technology that underpins all of the various technological flavours of 3G. Qualcomm has announced that it is having trouble meeting demand for W-CDMA radio chips. "As we get very near to seeing these things become a reality, we become more optimistic about what 3G can deliver," says Peter Bamford of Vodafone, the world's largest mobile operator. So now that it is finally happening, how does the reality of 3G stack up against the original vision?

Less data, more voice

That depends upon whom you ask. Mr Bamford, for example, denies that there has been any downgrading of the original vision. But he is at the most optimistic end of the spectrum, a view reflected in Vodafone's reluctance to write down the value of its 3G licences. Most observers agree that there has been a shift in expectations about how 3G networks will be used, away from video and other data services and towards traditional voice calling.

"In 2001, everyone was talking about video-telephony," says Mike Thelander of Signals Research Group, a consultancy. But while video-telephony sounds cool, the evidence from early 3G launches in Japan, South Korea, Britain and Italy is that hardly anybody uses it. Market research suggests that women are particularly reluctant to adopt it, says Mr Cole. Nokia's first mainstream 3G handset, the 7600, does not even support video calling, but nobody seems to mind.

Nor have the high hopes for data services been fulfilled – so far, at least. The idea was to encourage consumers to adopt data services on 2G phones, paving the way for fancier services on 3G phones. But while text-messaging is hugely popular, with over a billion messages sent daily worldwide, other forms of wireless data such as photo messaging, news updates, and music and game downloads have proved much less popular with consumers in most countries – Japan and South Korea are notable exceptions.

Such services "are still embryonic, but are going to be very important," insists Mr

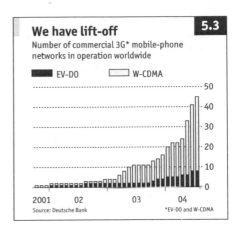

We have lift-off 5.3
Number of commercial 3G* mobile-phone networks in operation worldwide

■ EV-DO ☐ W-CDMA

Source: Deutsche Bank   *EV-DO and W-CDMA

Cole. Today's advanced handsets, he notes, are disrupting many industries simultaneously, including photography, music and gaming. The handset is slowly coming to be seen as "the Swiss Army knife of life services". But the changes will take years to play out, even though they are happening at breakneck speed. Mr Bamford likens the transformation in mobile phones over the past five years to the evolution of television over the past 40 years, from crude black-and-white to hundreds of digital channels in colour. "To expect customers to snap into this in five minutes is just unrealistic," he says.

Enthusiasm for data is growing, just not very fast: in September 2004 data services accounted for 16.3% of Vodafone's worldwide revenues, for example, up from 15% a year earlier. So hopes of a breakthrough in mobile-data usage still persist. At the moment, most optimism surrounds the prospects for music downloads to mobile phones (the most advanced models of which can now double as portable music players). Downloading ringtones is already popular, so downloading entire tracks – something that is only really practical using a 3G network – is the next logical step. Motorola, the world's second-largest handset-maker, has done a deal with Apple, whose iTunes Music Store dominates the market for legal music downloads. And Nokia has done a similar deal with LoudEye, another online music store. But it is still too early to tell whether this will turn into a mass market and, if it does, whether it will prove profitable for operators.

Greater emphasis is being placed instead on 3G's ability to deliver cheap voice calls – for as well as being able to support faster data downloads than 2G networks, 3G networks provide vast amounts of

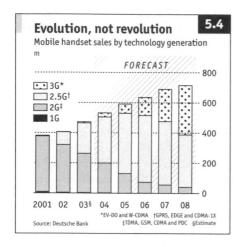

Evolution, not revolution `5.4`

Mobile handset sales by technology generation

m

FORECAST

800

600

400

200

0

- 3G*
- 2.5G†
- 2G‡
- 1G

2001 02 03§ 04 05 06 07 08

*EV-DO and W-CDMA †GPRS, EDGE and CDMA-1X
‡TDMA, GSM, CDMA and PDC §Estimate

Source: Deutsche Bank

voice capacity (typically three times as much as a 2G network) at a lower price (typically a quarter of the cost per minute). As a result, says Bob House, an analyst at Adventis, "operators' sights are now much more firmly trained on displacing voice from fixed networks."

By offering large bundles, or "buckets" of minutes as part of their monthly tariffs, operators hope to encourage subscribers to use their mobile phones instead of fixed-line phones, and even to "cut the cord" and get rid of their fixed-line phones altogether – something that is already happening, particularly among young people, in some parts of the world. In America, for example, where large bundles are commonplace, subscribers talk on their phones for 700 minutes per month on average, compared with 100 minutes per month in Europe, where call charges are much higher, notes Mark Heath of Analysys, a consultancy. Since 3G networks offer voice capacity at a quarter of the cost of 2G networks, it ought to be possible for operators to offer larger bundles at a lower price per minute and still make money.

But operators must price their bundles carefully, and distinguish between peak-time and off-peak minutes, to avoid getting caught out. Offering generous bundle deals may, for example, cannibalise revenues from their most valuable customers, who will quickly switch to a better deal. Operators also want to avoid having to spend money adding expensive base-stations to the busiest parts of their networks to handle peak load. And, of course, they want to avoid a price war. Although everyone agrees that the advent of 3G will cause the price of voice calls to fall and margins to decline, operators are in no hurry to cut their prices before they have to.

But there are signs that Hutchison 3G, a new operator that has launched 3G services in several European countries under the "3" brand, is already leading the European market down this path, notes Mr Thelander: in some cases, 3 offers voice calls for a fifth of the price of its rivals. Further pressure on pricing, argues João Baptista of Mercer Management Consulting, will come as fixed-line operators combat the flight of voice traffic to mobile with ultra-low-cost telephony services based on "voice over internet protocol" (VOIP) technology. With price cuts, he says, "someone starts, and then you can't stop it."

It would be a great irony if, after years of hype about data services, the "killer application" for 3G turned out to be boring old voice calls. In truth, however, nobody talks about killer apps any more. This reflects the realisation that 3G allows operators to offer lots of new services – music downloads, cheap voice calls, wireless broadband access to laptops – but that the appeal of these services will vary widely from one group of customers to another.

"Unlike traditional voice service, the adoption of 3G services is very much customer-segment specific," says Su-Yen Wong of Mercer. The lesson from Japan and South Korea, she says, is that "certain customer segments are interested in video, but others are not – some go for games, others for traffic updates." The challenge for 3G operators, she says, is to understand the appeal of different services to different types of customer.

The challenge of segmentation

That will require careful market segmentation. "3G gives you more scope, and segmentation broadly becomes more important," says Mr Bamford. The example of KTF, a South Korean operator, is instructive. It offers a service called Bigi Kiri to 13–18-year-olds (with unlimited text messaging between subscribers). Na, its brand for 18–25-year-old students, includes free cinema tickets and internet access at 68 universities; and Drama, another brand, caters for women. Other operators in South Korea and Japan do similar things.

The question for operators, says Mr Cole, is whether they can successfully appeal to all segments. At the moment, most operators have bland, generic brands that are intended to appeal to as broad a cross-section of the public as possible. But now they must decide whether to create sub-brands, or partner with other firms who are better able to appeal to specific demographic groups. There are already signs of this happening in many parts of the world as companies set themselves up as "mobile virtual network operators" (MVNOS).

Rather than build its own network, an MVNO teams up with an existing operator, and resells access to the operator's mobile network under its own brand. By far the best example is Virgin Mobile, an MVNO that resells airtime on T-Mobile's network in Britain, and Sprint's in America, to teenagers. The appeal for operators is that MVNOs enable them to reach out more effectively to customers. There has recently been a flurry of activity, with established brands including Tesco, 7-Eleven and MTV setting up as MVNOs.

Much of this activity has been prompted by the growing awareness that MVNOs are likely to have an important role in generating enough voice and data traffic to fill up those expensive new 3G networks. Since 3G phones can deliver graphics, music and video, large media firms, such as Disney, are actively investigating becoming MVNOs. Indeed, media giants might be more effective at driving uptake of data services than mobile operators, which are struggling to transform themselves from boring, technology-driven utilities into sexy consumer brands.

That in turn, suggests Mr Baptista, poses a long-term question for 3G operators: are they primarily network operators, or providers of services to consumers? No doubt some operators, with strong brands, will be able to hold their own against the likes of Disney. But second-tier operators might choose to focus on running a wholesale business, selling network capacity to others.

The calculations being made about the prospects for 3G are further complicated by the fact that the technology is still evolving, making new services possible. As things stand, the W-CDMA technology being adopted in much of the world has a maximum data-transfer rate of 384 kilobits per second. The rival 3G technology, called CDMA2000-1XEV-DO, which is already deployed in South Korea, Japan and parts of America, can deliver higher speeds of up to 2.4 megabits per second. In markets (such as Japan and America) where the two technologies compete side by side, W-CDMA operators are anxiously waiting for an upgraded version of the technology, called HSDPA, which will be faster still and was to make its debut in 2005.

Faster, better, sooner?

Never mind what all those letters stand for: the point is that as its speed and efficiency improves, 3G technology may, in some markets, start to compete with fixed broadband connections. Other, more obscure flavours of 3G technology, such as TDD-CDMA (again, never mind) and CDMA450 can also be used in this way. In New Zealand, Woosh Wireless is offering wireless broadband service using TDD-CDMA, while backers of CDMA450 point to its unusually long range, which makes it ideal for providing broadband in rural areas, as well as telephony. This opens up yet another new market for 3G operators.

3G is evolving in other ways, too. In 2003, SK, South Korea's leading mobile operator, launched a video-on-demand service over its 3G network. Subscribers paid a monthly fee of 20,000 won ($17) for access, and could then have movies beamed to their phones (while commuting, for example) for 1,000 won each. The service proved so popular that the 3G network could not cope, and SK had to raise its prices dramatically, causing demand to collapse. But evidently video does appeal to 3G subscribers, provided it is cheap enough. So SK has developed a hybrid satellite-cellular system. Some of its handsets have built-in satellite-TV receivers, offering 11 video and 25 audio channels. Meanwhile, both of the main 3G standards are being updated to allow for more efficient video broadcasts to handsets. Again, this could open new markets for 3G operators.

All of this makes it very difficult to answer the question of whether 3G will succeed, for 3G is a range of technologies that makes possible all kinds of new services. In Europe, 3G's main impact may simply be cheaper voice calls; in America, 3G may have most appeal to road warriors who want broadband access wherever they are; in the developing world, 3G could help to extend telephony and internet access into rural areas; and in South Korea and Japan, 3G might even – shock, horror – live up to the original lofty vision for the technology. The switching on of 3G networks around the world is not the end of the saga; the story continues to unfold.

The material on pages 162–9 first appeared in The Economist in September 2004.

Shape of phones to come

What is the best shape for a mobile handset – and what will the devices of the future look like?

IS YOURS A CANDY BAR, a clamshell, a slider, a jack-knife or a taco – and is it about to disappear, or break into several pieces? We are talking, of course, about mobile phones. Just a few years ago, they resembled bricks, but they now come in a baffling variety of shapes, sizes, colours and designs. This sudden proliferation of new handset shapes has been caused by the convergence of two trends: the mobile phone's growing importance as a fashion item, and advances in handset technology. Where will it all end?

The first company to realise what was going on was Nokia, the world's leading handset-maker. "We understood that the devices weren't technical devices any more but part of the end-user's personality," says Eero Miettinen, director of design at Nokia. In October 1999, the Finnish company set a trend when it launched its 8210 handset on the catwalk at fashion week in Paris. Design has since become an important way for handset-makers to make their products stand out. Siemens, for example, now sells a special collection of fashion handsets under the "Xelibri" brand.

The variety of handset designs has dramatically expanded as phones with colour screens and built-in cameras have become more popular. Around 70% of the handsets sold in 2004 had colour screens, up from 38% in 2003, and 44% had built-in cameras, up from 17% in 2003, according to UBS, an investment bank. The popularity of large colour screens has been driven by the growing adoption of data services such as web-browsing and downloading games and screen graphics. Furthermore, mobile phones can now be music players, photo albums and personal organisers too. The challenge of integrating all these new features into what was previously a voice-centric device has led to a flowering of innovative new designs.

In Japan and South Korea, where data services first took hold, the desire for a large screen in a small device led to the popularity of the clamshell, or "flip-phone" design. Allowing the phone to open and close protects the screen and also provides plenty of room for both display and keypad. Clamshells are now popular all over the world. Indeed,

Nokia's traditional preference for one-piece "candy bar" designs over clamshells has been widely blamed for the company's drop-off in sales in early 2004. Nokia responded by launching some new models, including several clamshells, and also cut prices to revitalise its sales.

But some observers believe Nokia may have lost its edge in design to its smaller rivals. Sony Ericsson, for example, has championed a clever new shape, called the swivel or jack-knife style, in camera-phones such as the S0505i, developed for the Japanese market, and the S700, for the rest of the world. When closed, the S700 resembles a camera, with a lens on the front and a large screen on the back. Its controls are deliberately modelled on those of a Sony digital camera. It can then be swivelled open, to reveal a standard mobile-phone keypad. "We believe form should follow function in a very classical way," says Hiroshi Nakaizumi, the head of Sony Ericsson's design centre.

Part of the appeal of the jack-knife design is that some Japanese consumers are getting bored with the clamshell design. But it is wrong to think that any one design will dominate in future, says Mr Nakaizumi. Instead, different types of users will want different styles, depending on whether they mainly use the devices for voice calls, text messaging, music or games. He suggests that in a few years' time, the market will roughly divide into three categories: traditional voice-centric handsets, "Swiss Army knife"-style phones that try to do everything (such as Sony Ericsson's own P900), and task-specific phones aimed at particular types of users, for whom telephony may be a secondary function.

Some such devices of this last type have already appeared, such as Nokia's N-Gage gaming console, the BlackBerry e-mail handheld (to which telephony functions have been added without changing its shape), and the slim digital camera that happens to be a phone as well, made by NEC for the Chinese market. The mobile phone could, in other words, be subsumed into another gizmo and disappear altogether, for some types of specialist users at least.

But an altogether more radical possibility is that the current "all in one" approach will give way to a more modular design, in which a basic handset is supplemented by add-ons connected via short-range "Bluetooth" wireless links. Owners of Bluetooth-capable phones can already make and receive calls with just a small headset or earpiece, while the handset sits in a nearby pocket, briefcase or handbag. The next step is to extend this approach to other devices. For example, a Bluetooth-equipped camera could send and receive pictures via a nearby handset, and a handheld gaming console could download new games and communicate with other players. The phone would act as a gateway between specialised local devices and the cellular network.

It is unclear whether or not this modular approach will appeal to consumers. The success of Apple's iPod, for example, which does one thing (playing music) and does it very well, may mean that task-specific phones will prove more popular than modular ones. And for users who want more functionality, the Swiss Army-knife approach has the advantage that you cannot leave bits of it behind, as you could with a modular phone. That said, the modular approach could make possible all kinds of radical designs, such as sunglasses or jewellery that also function as mobile headsets or display text messages.

There are already several examples of such techno-jewellery. The clip-shaped Xelibri 7 handset from Siemens, for example, is designed to be worn on the strap of a shoulder bag. "It looks familiar, but has a surprise built into it," says Leif Huff of IDEO, the firm that designed several of the Xelibri phones. Nokia, meanwhile, has launched a Bluetooth-compatible pendant containing a small screen. But while wearing a wireless headset is starting to become more socially acceptable, wearing your phone is still considered geeky, Mr Huff observes.

What is clear is that the mobile handset is now much more than just a phone, and depending on what else you want to do with it, it may assume a very different shape altogether. It may even need a new name. Indeed, at Motorola, the second-biggest handset-maker, the term "cell phone" has now been banned. The handset is now turning into more of a "personal network device" says Tom Lynch, president of Motorola's handset business. "We are trying to think more broadly about it," he says, "which is why we call it 'the device formerly known as the cell phone'."

The material on pages 170–2 first appeared in *The Economist* in June 2004.

The origins of Vertu

Is there really a market for a $20,000 mobile phone?

IF YOU CAN SPEND $20,000 on a watch, why not on a mobile phone? That is the idea behind Vertu, which describes itself as "the first luxury communication company". Its stylish handset – the company prefers to call it an "instrument" – features a sapphire-crystal screen and ruby bearings, and is available in stainless steel, gold and platinum finishes, with prices ranging from $4,900 to $19,450. Since its launch, the phone – sorry, instrument – has become a celebrity favourite. Gwyneth Paltrow, an American actress, was the first customer. Madonna and Mariah Carey are said to be Vertu fans; another singer, Jennifer Lopez, is reported to own three. Vertu is the brainchild of Frank Nuovo, a design guru at Nokia, the world's largest handset-maker, of which Vertu is a subsidiary. But is there really a big enough market for luxury phones?

The company insists there is. After all, as with luxury watches, inkpens and cars, your choice of mobile phone is increasingly a form of self-expression. And although celebrity customers get all the attention, says Danielle Keighery of Vertu, obscure rich people are buying the phones too. She refuses to be drawn on sales figures, other than to say that the firm is "very pleased" with the response since the handsets went on sale in 2002.

A big selling-point is a special button that connects the user to Vertu's dedicated concierge service, which can organise travel, restaurant and hotel bookings, or find a good doctor or florist in a foreign city. When Ms Paltrow mislaid her phone's charger she called the concierge and a new one arrived within minutes. The concierge service is available worldwide in five languages. Detailed records are kept of each customer's preferences.

By selling handsets on the basis of craftsmanship, style and service, rather than whizzy features, Vertu is taking a different approach to that of the technology-obsessed mobile-telephone industry. It prefers to launch its products at fashion shows rather than industry conferences. The Vertu handset's monochrome screen, above-average weight, and lack of support for such technologies as Bluetooth and GPRS leaves geeks unimpressed. But they are not the target market. Besides, new features can be added by switching the handset's removable innards.

Not everybody is convinced. Sagra Maceira de Rosen, a luxury-goods analyst at J.P. Morgan, thinks that Vertu is aiming too high. The idle rich who are expected to buy the phones are unlikely to need the concierge service because they have armies of assistants already, she suggests. The firm should perhaps be aiming at busy investment bankers instead (if it can find any). Nor is Vertu a logical brand-building exercise for Nokia, observes Ben Wood, an analyst at Gartner. Vertu's parentage is kept quiet, so Nokia's mass-market handsets are unlikely to benefit by association. Nokia already has by far the strongest brand in the industry.

But many luxury brands, observes Ms Keighery, subsequently launch more affordable versions of their products. So the gap between Vertu's cheapest phone and Nokia's most expensive may yet be closed. In the longer term, Vertu plans to exploit the emergence of "wearable" technology, as phones morph into jewellery.

Here, Vertu may be on to something, says Sofia Ghachem, an analyst at UBS Warburg. Siemens, another handset-maker, has launched a range of wearable "fashion accessory phones" under the name Xelibri, in the hope that marketing phones as fashion items will encourage people to buy new handsets more often. With market penetration at around 85% in western Europe, growth in handset sales is slowing and Siemens believes its new approach could give the industry a much-needed boost.

The industry's official line is that future growth will come from the adoption of whizzy new data services delivered over third-generation networks. But demand for such services is still uncertain. Pushing phones explicitly as fashion accessories, as Vertu and Siemens are doing, might be a good fall-back plan. If data services turn out to be a niche market, says Ms Ghachem, "fashion is another way to go".

The material on pages 173-4 first appeared in The Economist in February 2003.

Why phones are the new cars

And why this is a good thing

"PARKS BEAUTIFULLY", boasts an advertising hoarding for the XDA II, above a glimpse of its sleek silver lines. "Responsive to every turn", declares another poster. Yet these ads, seen in London in 2004, are selling not a car, but an advanced kind of mobile phone. Maybe that should not be a surprise. Using automotive imagery to sell a handset makes a lot of sense for, in many respects, mobile phones are replacing cars.

Phones are the dominant technology with which young people, and urban youth in particular, now define themselves. What sort of phone you carry and how you customise it says a great deal about you, just as the choice of car did for a previous generation. In today's congested cities, you can no longer make a statement by pulling up outside a bar in a particular kind of car. Instead, you make a similar statement by displaying your mobile phone, with its carefully chosen ringtone, screen logo and slip cover. Mobile phones, like cars, are fashion items: in both cases, people buy new ones far more often than is actually necessary. Both are social technologies that bring people together; for teenagers, both act as symbols of independence. And cars and phones alike promote freedom and mobility, with unexpected social consequences.

The design of both cars and phones started off being defined by something that was no longer there. Cars were originally horseless carriages, and early models looked suitably carriage-like; only later did car designers realise that cars could be almost any shape they wanted to make them. Similarly, mobile phones used to look much like the pushbutton type of fixed-line phones, only without the wire. But now they come in a bewildering range of strange shapes and sizes.

Less visibly, as the structure of the mobile-phone industry changes, it increasingly resembles that of the car industry. Handset-makers, like carmakers, build some models themselves and outsource the design and manufacturing of others. Specialist firms supply particular subassemblies in both industries. Outwardly different products are built on a handful of common underlying "platforms" in both industries, to reduce costs. In each case, branding and design are becoming more important as the underlying technology becomes more interchangeable.

In phones, as previously happened in cars, established western companies are facing stiff competition from nimbler Asian firms. Small wonder then that Nokia, the world's largest handset-maker, recruited its design chief, Frank Nuovo, from BMW.

That mobile phones are taking on many of the social functions of cars is to be welcomed. While it is a laudable goal that everyone on earth should someday have a mobile phone, cars' ubiquity produces mixed feelings. They are a horribly inefficient mode of transport – why move a ton of metal around in order to transport a few bags of groceries? – and they cause pollution, in the form of particulates and nasty gases. A chirping handset is a much greener form of self-expression than an old banger. It may irritate but it is safe. In the hands of a drunk driver, a car becomes a deadly weapon. That is not true of a phone. Despite concern that radiation from phones and masts causes health problems, there is no clear evidence of harm, and similar worries about power lines and computer screens proved unfounded. Less pollution, less traffic, fewer alcohol-related deaths and injuries: the switch from cars to phones cannot happen soon enough.

The material on pages 175–6 first appeared in *The Economist* in April 2004.

Think before you talk

Can technology make mobile phones less socially disruptive?

THE MOBILE PHONE IS A PARADOXICAL DEVICE. Its primary function is social: to enable its owner to communicate with other people. At the same time, though, using a mobile phone can seem profoundly anti-social, not least to people in the immediate vicinity. In restaurants, theatres and museums, on trains, or even standing in the supermarket checkout queue, there is no escape from chirping and bleeping phones, nor from the inane conversations of their owners. In 2002 Philip Reed, a New York councillor, proposed a law that would prohibit the use of mobile phones in "places of public performance", such as theatres, art galleries and concert halls, punishable by a $50 fine. But his proposal was derided as unenforceable. Might a technological approach to taming the mobile phone, and the behaviour of its users, be more successful?

Crispin Jones, Graham Pullin and their colleagues at IDEO, an industrial-design company, think the answer is yes. (IDEO is responsible for designing such products as the Palm V pocket computer, the original Microsoft mouse, the TiVo personal video-recorder and the world's most high-tech dressing rooms, at Prada in New York.) As part of an internal research project, the team designed five prototype "social mobiles" which modify their users' behaviour to make it less disruptive.

For example, the first phone, called SOMO1, gives its user a mild electric shock, depending on how loudly the person at the other end is speaking. This encourages both parties to speak more quietly, otherwise the mild tingling becomes an unpleasant jolt. Such phones, the designers suggest archly, could be given to repeat offenders who persistently disturb people with intrusive phone conversations.

SOMO2 is a phone intended for use in situations (such as a hushed art gallery) where speaking is inappropriate. Manipulating a joystick and a pair of saxophone keys controls a speech synthesiser that produces an expressive range of vowel sounds for non-verbal communication: "Hmm? Yeah." The third phone, SOMO3, resembles a small, clarinet-like musical instrument. Dialling is done by holding down combinations of keys and blowing; tunes replace phone numbers. "The public performance that dialling demands acts as a litmus test of when it is appropriate to make a call," say the designers.

somo4 replaces ringtones with a knocking sound: to make a call, select the number and knock on the back of the phone, as you would on somebody's door. The recipient of the call hears this knock (cleverly encoded and relayed via a short text-message) and decides how urgent the call is. How you knock on a door, says Mr Pullin, is freighted with meaning: there is a world of difference between tentative tapping and insistent hammering. somo5 has a catapult-like device that can be used to trigger intrusive sounds on a nearby user's phone, anonymously alerting them that they are speaking too loudly.

None of these phones is intended as a commercial product; the design team simply hopes to provoke discussion. It seems to be working. The project won a prize from the Agency of Cultural Affairs in Japan, perhaps the country where both social etiquette and mobile phones are taken more seriously than anywhere else. And behind these silly-sounding phones is a serious point. Much is made of "user-centric" design, says Mr Pullin, but in the case of mobile phones, the people surrounding the user need to be considered too.

The material on pages 177–8 first appeared in *The Economist* in January 2003.

Move over, Big Brother

Privacy advocates have long warned of states spying on citizens. But technology is, in fact, democratising surveillance

LIVING WITHOUT PRIVACY, even in his bedroom, was no problem for Louis XIV. In fact, it was a way for the French king to demonstrate his absolute authority over even the most powerful members of the aristocracy. Each morning, they gathered to see the Sun King get up, pray, perform his bodily functions, choose his wig and so on. One reported in 1667 that there "is no finer sight in the world than the court at the *lever* of the King. When I attended it yesterday, there were three rooms full of people of quality, such a crowd that you would not believe how difficult it was to get into His Majesty's bedchamber."

Will this past – life without privacy – be our future? Many futurists, science-fiction writers and privacy advocates believe so. Big Brother, they have long warned, is watching. Closed-circuit television cameras, which are proliferating around the world, often track your moves; your mobile phone reveals your location; your transit pass and credit cards leave digital trails. "Light is going to shine into nearly every corner of our lives," wrote David Brin in his 1998 book *The Transparent Society* (Perseus Publishing). The issue, he argued, is no longer how to prevent the spread of surveillance technology, but how to live in a world in which there is always the possibility that citizens are being watched.

But in the past few years, something strange has happened. Thanks to the spread of mobile phones, digital cameras and the internet, surveillance technology that was once mostly the province of the state has become far more widely available. "A lot has been written about the dangers of increased government surveillance, but we also need to be aware of the potential for more pedestrian forms of surveillance," notes Bruce Schneier, a security guru. He argues that a combination of forces – the miniaturisation of surveillance technologies, the falling price of digital storage and ever more sophisticated systems able to sort through large amounts of information – means that "surveillance abilities that used to be limited to governments are now, or soon will be, in the hands of everyone".

Digital technologies, such as camera-phones and the internet, are very different from their analogue counterparts. A digital image, unlike

179

a conventional photograph, can be quickly and easily copied and distributed around the world. (Indeed, it is easier to e-mail a digital image than it is to print one.) Another important difference is that digital devices are far more widespread. Few people carry film cameras with them at all times. But it is now quite difficult to buy a mobile phone without a built-in camera – and most people take their phones with them everywhere. According to IDC, a market-research firm, 264m camera-phones were sold in 2004, far more than film-based cameras (47m units) or digital cameras (69m units) combined.

The speed and ubiquity of digital cameras lets them do things that film-based cameras could not. For example, the victim of a robbery in Nashville, Tennessee, used his camera-phone to take pictures of the thief and his getaway vehicle. The images were shown to the police, who broadcast descriptions of the man and his truck, leading to his arrest ten minutes later. Other similar stories abound: in Italy, a shopkeeper sent a picture of two men who were acting suspiciously to the police, who identified them as wanted men and arrested them soon afterwards, while in Sweden, a teenager was photographed while holding up a corner shop, and was apprehended within an hour.

Watching your every move

The democratisation of surveillance is a mixed blessing, however. Camera-phones have led to voyeurism – and new legislation to strengthen people's rights to their own image. In September 2004, America's Congress passed the Video Voyeurism Prevention Act, which prohibits the photography of various parts of people's unclothed bodies or undergarments without their consent. The legislation was prompted both by the spread of camera-phones and the growing incidence of hidden cameras in bedrooms, public showers, toilets and locker rooms. Similarly, Germany's parliament has passed a bill that outlaws unauthorised photos within buildings. In Saudi Arabia, the import and sale of camera-phones has been banned, and religious authorities have denounced them for "spreading obscenity". A wedding in the country turned into a brawl when one guest started taking pictures with her phone. South Korea's government has ordered manufacturers to design new phones so that they beep when taking a picture.

There are also concerns about the use of digital cameras and camera-phones for industrial espionage. Sprint, an American mobile operator, is now offering one of its bestselling phones without a camera in response to demands from its corporate customers, many of which have banned cameras in their workplaces. Some firms make visitors and staff leave camera-phones at the entrance of research and manufacturing facilities – including Samsung, the South Korean company that pioneered the camera-phone.

Cheap surveillance technology facilitates other sorts of crime. Two employees at a petrol station in British Columbia, for example, installed a hidden camera in the ceiling above a card reader, and recorded the personal identification numbers of thousands of people. They also installed a device to "skim" account details from users as they swiped their plastic cards. The two men gathered the account details of over 6,000 people and forged 1,000 bank cards before being caught.

In another case, a man installed keystroke-logging software, which monitors every key pressed on a computer's keyboard, on PCs in several Kinko's copy shops in New York City. (Keystroke-logging software is sold for use by businesses to monitor their employees, or by parents who wish to monitor their children's activities online.) This enabled him to remotely capture account numbers and passwords from over 450 people who rented the terminals, and to siphon money out of their bank accounts.

Surveillance is a two-way street

But the spread of surveillance technology also has its benefits. In particular, it can enhance transparency and accountability. More and more video cameras can be found in schools, for example. Web-based services such as ParentWatch.com and KinderCam.com link to cameras in hundreds of American child-care centres, so that parents can see what their offspring (and those looking after them) are up to. Schools are also putting webcams in their classrooms: one American school district has planned to install 15,000 such devices for use by security personnel (and, perhaps one day, parents). And tech firms such as Google have put webcams in their staff restaurants, so employees can delay going to lunch if they see a long queue.

Steve Mann, a professor at the University of Toronto, calls the spread of citizen surveillance "sousveillance" – because most cameras no longer watch from above, but from eye level. Instead of being on top of buildings and attached to room ceilings, cameras are now carried by ordinary people. The video images of Rodney King being assaulted by police officers and the horrific pictures of prisoner abuse from the Abu Ghraib jail in Iraq are the best known examples. But as Mr Mann and his colleagues organised the first "International Workshop on Inverse Surveillance" in

April 2004, there was no shortage of reports on other cases: in Kuwait, a worker took photos of coffins of American soldiers being loaded on to a plane; in New Jersey, a teenager scared off a kidnapper by taking his picture; in Strasbourg, a member of the European Parliament filmed colleagues making use of generous perks.

Camera-phones could have a profound effect on the news media. Technologies such as newsgroups, weblogs and "wikis" (in essence, web pages which anybody can edit) let people distribute images themselves, bypassing the traditional media, notes Dan Gillmor, a journalist, in his book *We the Media* (O'Reilly, 2004). Camera-phones make everyone a potential news photographer. Unsurprisingly, old media is starting to embrace the trend. The *San Diego Union-Tribune* launched a website to gather camera-phone images of news events taken by its readers, and the BBC also encourages users of its website to send in pictures of news events.

Companies and governments will have to assume that there could be a camera or a microphone everywhere, all the time, argues Paul Saffo of the Institute for the Future. Unsafe conditions in a factory or pollution at a chemical plant are harder to deny if they are not just described, but shown in photos and videos. Animal-rights activists, for instance, operate online multimedia archives where people can store and view graphic images from chicken farms, slaughterhouses and fur factories. Such material can cause outrage among consumers, as was the case with videos of dolphins caught in tuna nets.

In 2003, a German member of parliament was caught photographing a confidential document of which only a few copies were handed out (and later collected) at a background meeting on health-care reform. Some Berlin politicians are said to let reporters eavesdrop on fellow parliamentarians by calling them just before an important meeting and then failing to hang up, in effect turning their phones into bugs.

In November 1996, Senegal's interior minister was caught out when he admitted that there had been fraud in a local election, but failed to notice that a bystander was holding a mobile phone with an open line. The election was annulled. In the same country's presidential election in 2000, radio stations sent reporters to polling stations and equipped them with mobile phones. The reporters called in the results as they were announced in each district, and they were immediately broadcast on air. This reduced the scope for electoral fraud and led to a smooth transfer of power, as the outgoing president quickly conceded defeat.

The social consequences of the spread of surveillance technology remain unclear. Mr Brin suggests that it could turn out to be self-regulating: after all, Peeping Toms are not very popular. In a restaurant it is generally more embarrassing to be caught staring than to be observed with crumbs in your beard. "A photographically 'armed' society could turn out to be more polite," he suggests, referring to an American aphorism that holds "an armed society is a polite society". Alternatively, the omnipresence of cameras and other surveillance technologies might end up making individuals more conformist, says Mr Brin, as they suppress their individuality to avoid drawing too much attention to themselves.

The surveillance society is on its way, just as privacy advocates have long warned. But it has not taken quite the form they imagined. Increasingly, it is not just Big Brother who is watching – but lots of little brothers, too.

The material on pages 179–83 first appeared in *The Economist* in December 2004.

6
GAMING

Gaming goes to Hollywood

The games business is becoming more like Hollywood, but differences remain

THE LATEST JAMES BOND ADVENTURE is at number one, *Finding Nemo* at number 12, and *The Lord of the Rings* at number 18. A chart of DVD sales or box-office receipts? No, this is the British video-games chart in March 2004, a vivid illustration of how the once-separate worlds of movies and games have become increasingly intertwined. Film actors, even famous ones, now voice their characters in games too. Animators, artists and model-makers move freely between the two worlds; the same companies produce trailers both for games and for films. People in the booming games business are fond of pointing out that worldwide games sales, at around $20 billion a year, now exceed movie box-office revenues. With every passing day, gaming seems less like Hollywood's poor relation.

The convergence between films and games makes sense for many reasons. Both special-effects-laden blockbusters and shoot-'em-ups rely on computer power, and as games consoles become more capable their output becomes ever more cinematic. Indeed, modern games based on *Star Wars* look even better than the original films, since today's games consoles far outperform any special-effects technology available back in the 1980s. Costs have increased as the production values of games have improved: the typical budget is now $5m–8m.

Where tie-in games used to be an afterthought, they are now integrated into the film-making process from the start, says Robert Kotick, the boss of Hollywood-based Activision, the second-largest games publisher. "Studios used to regard game licensees like the T-shirt licensees," he says, and only provide access to artwork late in the film-making process. But no longer. His firm had a dozen people inside the studios of DreamWorks working on the game of *Shrek* 2, an animated film released in May 2004.

Creativity killer

The worry, however, is that gaming is also becoming Hollywood-like in a less desirable sense. Nick Gibson of Games Investor, a consultancy, notes that as costs rise, risk-averse games publishers regard film franchises as a safe bet. But as movie tie-ins and sequels proliferate, there is concern that creativity is suffering. "We are seeing less and less innovation, because the stakes are so high," says Scott Orr, a former executive at Electronic Arts, the biggest games publisher. He founded Sorrent, a firm that makes games for mobile phones. With their small screens and limited processing power, mobile phones are a throwback to the 1980s, when a lone programmer toiling over a computer could cook up a hit game in a few weeks. "It allows us to innovate again," says Mr Orr.

It is also getting harder for independent developers to secure publishing deals, particularly as the big publishers take more development in-house. But, says Mr Gibson, small developers can still do well if their games are of sufficient quality. A good example is Pivotal, a British company based in a converted barn near Bath, and maker of a series of hit military games.

For his part, Mr Kotick insists that the growing scale and maturity of the games industry is not stifling creativity. Indeed, he says, it is important not to take the comparison with Hollywood too far. He moved his company from San Francisco to Hollywood in 1992, expecting to be able to exploit synergies between the two industries, but found that proximity to Hollywood was of limited value.

For despite the convergence between games and films, the games industry is still different in a number of important respects. For instance, game-making is iterative: it is possible to market-test a game and then modify it, and to do so repeatedly during the game's development cycle – something that is not really possible with a film. That makes the games business far more predictable, and accounts for its far higher returns. Activision's game *True Crime*, for example, released in November 2003, raked in $100m in its first month. Only two films released that month earned as much, says Mr Kotick, and they cost far more to make.

Another notable difference, says Mr Gibson, is that in the games world, sequels often outsell the original game – something that is very unusual in the movie business. That is because, while the plots and jokes of movie sequels are mostly thinner than the original, there is scope to improve on a successful game with technological enhancements (such as better graphics) and features such as new weapons, vehicles or character abilities. These ensure that many game sequels are better than the original. As a result, gaming concepts and brands are extremely valuable.

Much has been made of the trend to sell the movie rights to popular games. The best known example is *Tomb Raider*, though the most recent game in the series, and the second of two films based on the franchise, were flops. But even though the film rights to many games have been sold – such as those to *Prince of Persia* and *Soul Calibur* – the main trend will continue to be to make games from films, not vice versa. Indeed, says Mr Kotick, publishers are starting to think twice about selling the rights to their games, since a bad film adaptation – over which the game publisher has little control – can tarnish a lucrative game franchise. There could turn out to be limits to the cosiness between the two industries after all.

The material on pages 186–8 first appeared in *The Economist* in March 2004.

The Halo effect

Games sales exceed movie box-office receipts. But are the two comparable?

THERE IS NOTHING THE VIDEO-GAME INDUSTRY likes more than boasting about how it is bigger than the film business. In November 2004, for the launch of *Halo 2*, a shoot-'em-up that runs on Microsoft's Xbox console, over 6,000 shops across America opened their doors at midnight to sell the game to queues of fans. Add over 1.5m pre-orders and the game brought in over $100m in its first day on sale. Bill Gates, Microsoft's chairman, called it "an opening day that's greater than any motion picture has ever had in history". For example, *The Incredibles*, a blockbuster film from Pixar, took a mere $70.5m in its opening weekend, while the record for an opening day's ticket sales, at $40.4m, is held by *Spider-Man 2*. Overall, annual game sales, at around $20 billion, now exceed box-office receipts.

But the two are not really comparable. Film-going is mainstream: nearly everybody does it at some time. Around 10m people saw *The Incredibles* in its opening weekend, and perhaps 50m people will see it in cinemas eventually. Even more will view it on television, DVD, or on a plane. Playing video games is still a minority sport, though its popularity has soared. Not everybody wants a games console. It is only because games cost so much more than film tickets ($50 versus $7) that games can outsell films, despite their narrower appeal.

Games and films would be more comparable if more people played games, there was no need to buy a console and individual games cost as little as film tickets. All of that describes mobile-phone gaming. Having moved from the bedroom to the living room, gaming is now moving on to the mobile handset, says Brian Greasley of I-play, a mobile-games firm.

According to the 2004 Mobinet study of 4,500 mobile users in 13 countries by A.T. Kearney, a consultancy, and the Judge Institute of Management at Cambridge University, the number of people who download games to their phones grew in 2004 to 10% of the world's 1.7 billion mobile users, exceeding the number of console users.

The fragmented mobile-games industry is consolidating fast, and there are more higher-quality games based on big franchises such as

Spider-Man. Jamdat, a mobile-games firm, went public in October 2004. Most tellingly of all, the big boys are moving in. "There is the potential for a massive market in the future," says Bruce McMillan of Electronic Arts, the world's biggest games publisher, "so it makes sense for us to have a stronger presence." Yet Screen Digest, a consultancy, reckons that the value of the mobile-gaming market only exceeded $1 billion for the first time in 2004. Small change next to the film business – but, perhaps, the more valid comparison.

The material on pages 189–90 first appeared in *The Economist* in November 2004.

Hand-to-hand combat

Sony takes on Nintendo again, this time with a new portable games console

IN DECEMBER 1994, Sony shook up the cosy world of video gaming, then dominated by Nintendo. Rather than appeal to the young teenagers who were Nintendo's main customers, Sony aimed its new PlayStation console at an older audience: late teens and twentysomethings. The games were darker, more sophisticated, and often more violent. Sony correctly identified a "PlayStation generation" that had grown up with gaming and wanted to keep playing beyond their teens. This hugely boosted the market, since older gamers (today, the average age of a console owner is around 28) have more disposable income. Now, having shown that gaming on fixed consoles (which plug into televisions) is not just for kids, Sony hopes to do the same for portable gaming, with the launch in Japan in December 2004 of the PlayStation Portable, or PSP, a hand-held gaming device.

Once again, it is attacking a market dominated by Nintendo, which has sold over 150m of its Game Boy hand-helds since 1989 and still has a market share of over 90% in the hand-held market, despite having lost control of the fixed-console market to Sony. And once again there is a clear opportunity to expand the market, since most games available for the Game Boy are developed by Nintendo itself and have little appeal to anyone over 16.

That is why Sony's PSP is aimed at older gamers, aged 18–34. Its range of games is more varied and sophisticated than the Game Boy's, thanks to Sony's established relationships with game developers. It can double as a music and video player, and Sony has hinted at future phone and camera attachments too. This versatility should make the PSP more appealing to older users, says Nick Gibson of Games Investor. Moreover, he notes, the price (¥19,800, or around $190) is low enough to be within the reach of younger gamers – the result, claims Sony, of making many of the PSP's parts itself. By pitching the PSP at a demographic between the Game Boy and the PlayStation, Sony is hoping both to steal market share from Nintendo and attract older users.

Nintendo's response was a pre-emptive strike: it launched its own new hand-held console, the Nintendo DS, in America and Japan in

November and December 2004 respectively. The DS (which takes its name from its innovative "dual screen" design) is aimed at gamers aged 17–25. Compared with the Game Boy, which will continue to cater for younger users, the DS has a more varied range of games: Nintendo has worked hard to get third-party software publishers to support the device. Its marketing slogan, "Touching is good", a reference to the unit's touch-screen, is uncharacteristically risqué for Nintendo. The DS has been a smash hit: 500,000 units were sold in the first week in America alone. Nintendo expected to sell a total of 2.8m units in America and Japan by the end of 2004, and launched the DS in Europe in March 2005. Apple's iconic iPod music player, Nintendo points out, took 19 months to sell 1m units.

The launch of the PSP in Japan (and in Europe and America in 2005) puts the two new hand-helds in direct competition for the first time. But the PSP has a lot of catching up to do, since Sony had only 200,000 units available at the launch, all of which sold almost immediately. Sony planned to ramp up production in 2005, and expected to have sold 3m units by the end of March. In the longer term, however, "both can succeed," says Brian O'Rourke, an analyst at In-Stat/MDR, a market-research firm. To some extent they are aimed at different audiences: the DS is more likely to appeal to Nintendo's existing customers, and the PSP to Sony's.

Yet despite the early lead established by the DS, Sony seems certain to loosen Nintendo's longstanding grip on the hand-held market, at least to some extent. But by how much? One crucial factor is the extent to which third-party games developers support the new devices. Historically, notes Mr Gibson, they have shunned hand-helds, partly because they were reluctant to compete with Nintendo in developing Game Boy games, and also because margins on hand-held games are lower than on fixed-console games.

By introducing competition and exploiting its close existing relationships with developers, Sony is hoping to make the hand-held market more closely resemble the fixed-console market, which it dominates. If that were not enough, it also hopes to establish the PSP as the "Walkman of the 21st century", the first of a new family of products that will meld gaming with music and video playback on the move. The Walkman and the PlayStation, Sony's most successful products, are certainly illustrious parents. But that is no guarantee of success. And having lost out to Sony in the fixed-console market, Nintendo is putting up quite a fight to defend its remaining stronghold.

POSTSCRIPT

By the end of March 2005, Nintendo had sold over 5m DS consoles, and Sony had sold 2.5m PSPs, missing its target of 3m. The European launch of the PSP was delayed until September, in part due to stronger than expected demand in America. With each of the two consoles selling well, there would indeed seem to be room in the market for both of them.

The material on pages 191–3 first appeared in *The Economist* in December 2004.

Playing to win

How close is the relationship between real-world skills and video games, on playing fields and battlefields?

IS DWIGHT FREENEY the best player in America's National Football League (NFL)? He has a clearer grasp than anyone else of the strategy and tactics necessary to win in the brutal, chaotic game. He also has the quick reactions necessary to respond to the rapidly changing conditions at the line of scrimmage, where nimble 150kg giants, bulging with padding and sporting helmets and face-masks, barrel into and around each other. How can you tell? He won the 2004 "Madden Bowl", a video-game tournament in which NFL players compete on screen, rather than on the field, and which is held each year just before the Super Bowl, the championship of real-world American football. On the screen, Mr Freeney dominated, holding his opponents scoreless.

On the real-world field too, Mr Freeney is by all accounts a good player. A defensive end, he is unusually fast and can put a great deal of pressure on the opposing team's quarterback. However, in the 2003–04 season, when Mr Freeney won the Madden Bowl, the defensive line for his team, the Indianapolis Colts, was one of the worst in the league, according to footballoutsiders.com, a website which performs rigorous statistical analyses of the game of football. At the end of 2004, the Colts' defensive line was ranked dead last. Other football players who did well in the virtual championship – such as David Carr, a mid-ranked quarterback for the Houston Texans, and Dante Hall, a mediocre wide receiver for the Kansas City Chiefs – have similarly unremarkable records in real life.

Mr Freeney demonstrates that real-world skills translate readily into the virtual world: professional football players turn out to be good at virtual football, too. But what if the skills do not translate in the other direction? The assumption that skills learned in a simulated environment can be readily transferred into the real world is widespread in fields including pilot training and, increasingly, military training. But is it correct?

Lock and load

As video-game technology has steadily improved and the gadgets of war have grown more expensive, America's military is relying more heavily on computer games as training tools. Some games which the military uses are off-the-shelf products, while others are expensive, proprietary simulations. A 2001 report by RAND, a think-tank, boosted the enthusiasm for military gaming when it concluded that the middle ranks of the army were experiencing a "tactical gap". Because most lieutenants and captains had not commanded troops in battle, or had not trained extensively enough in mock battles, they lacked the know-how necessary to do their jobs well. Fixing this, either by keeping infantry commanders in their jobs longer or by stepping up the pace of training, proved difficult – which led to a proliferation of initiatives in different branches of the military to develop games for training purposes.

The "tactical gap" may now have disappeared, as a result of the war in Iraq. A paper published in the summer of 2004 by Leonard Wong of the Army War College in Carlisle, Pennsylvania, asserted that the "complexity, unpredictability and ambiguity of post-war Iraq is producing a cohort of innovative, confident and adaptable junior officers". Nonetheless, games remain a far cheaper training method than invading countries and waging wars. Yet their true effectiveness is far from certain. An eagerness on the part of the military to save money and embrace a transformative mission, and an eagerness on the part of the gaming community to see itself as genuinely useful, rather than as merely providing frivolous entertainment, may be obscuring the real answers.

In the case of football, there is no shortage of data to analyse. Not only is there a score at the end, and a clear winner and loser, but a multitude of data can be harvested as the game is under way – passes completed, sacks allowed, fumbles forced and so on. For those unfamiliar with American football, the details of these data are unimportant – the relevant fact is that they exist. The same cannot be said on a battlefield. In the proverbial fog of war, there is no easy way quantitatively to measure success or failure in the many different aspects of warfare.

Other sports, especially baseball, offer a greater wealth of data. However, no other sport seems to match the set of psychological and physical skills needed on a battlefield so well. Vince Lombardi, probably the most famous coach in American football's history, enjoyed comparing the football field to a battlefield. But the more important comparison is the converse – that a battlefield can seem like a football field, according to Lieutenant-Colonel James Riley, chief of tactics at the Army Infantry School in Fort Benning, Georgia. Indeed, Colonel Riley says his commanding general makes this very analogy constantly. In football, as in infantry combat, a player must be aware of both the wider situation on the field, and the area immediately surrounding him. The situation changes rapidly and the enemy is always adapting his tactics. Physical injuries abound in both places. Football is as close to fighting a war as you can come without guns and explosives.

The generals would thus be chagrined to hear Mr Freeney say that while playing football has made him better at the video game, the video game has not affected his real-world performance. Mr Freeney highlights the surreal experience of playing a video game where he knows the onscreen characters (EA Sports, the manufacturer of Madden, is proud of its realistic depictions of real-world football players) and indeed of playing as himself on screen. "It's not the total Dwight Freeney," he says. "There are some similarities." For the military, which is training soldiers for life-or-death situations, are "some similarities" enough?

According to Colonel Riley, they just might be. All training exercises, whether in a mocked-up urban combat environment or on a computer screen are, he says, "partial task simulators". The army will not, after all, actually try to get its soldiers to kill each other for practice. And Colonel Riley asserts that some games, in particular *Full Spectrum Command*, a game he uses to train infantry captains, can usefully impart a partial skill set. The single most important thing for a simulation to achieve, he says, is the suspension of disbelief. This is easy to achieve in, say, a flight simulator. When flying a real aircraft, the pilot sits in a seat and manipulates controls, looking at a screen – much as he does in a simulator.

But simulating infantry combat, as Colonel Riley is doing, is much more difficult. As he admits, he is not certain how much "simulation dexterity translates into reality". However, he maintains that *Command* is a useful training tool. An infantry captain commands a company of 130 men. Without the simulation, putting a new captain through an exercise meant using 129 men as training tools – an enormous overhead. Colonel Riley says that the simulation, however flawed, is an improvement. It can help to teach a captain battlefield tactics – how to deploy troops, when to call in artillery or airstrikes, and so on. And Colonel Riley says that the game has sufficient fidelity to the real world: the graphics are good enough, and the artificial intelligence of the enemy clever enough, to help teach captains how to make command decisions.

Paradoxically, the larger the scale of the situation being simulated, the better and more useful a simulation might be. *Full Spectrum Warrior*, a game which, like *Command*, was developed under the auspices of the

Institute for Creative Technologies (ICT), a military-funded centre at the University of Southern California, is a "first-person shooter" game which simulates infantry combat at the squad level – ten individuals or so. It has received a lot of attention because it exists in two versions – one of which is used as a military training tool, while the other is on sale to the public.

The ICT trumpets it as an especially accurate rendition of close infantry combat, developed in co-operation with the infantry school at Fort Benning. However, Colonel Riley says that it does not meet the needs of any of his courses, and that when infantrymen play the game, they complain of its lack of fidelity. The smaller the simulation, the bigger the disjunction between the tasks necessary in reality and those on the computer.

The opposite extreme is exemplified by OneSaf, a large-scale simulation being developed by the army in Orlando, Florida. The goal of OneSaf is extremely ambitious: to simulate the entire army. Unlike *Warrior* and *Command*, it is not meant to be used as a training tool to hone soldiers' instincts, but by planners and, in the long run, even front-line troops, to see what would happen in a given situation. OneSaf is an enormously complicated software framework which is expected to take years to develop.

It will surely not be a perfect recreation of the world. But it illustrates the power of technology to be transformative in a way that *Warrior* and *Command* are not – since they are not as good as training exercises in the field, but merely a cheaper alternative. OneSaf, if it works, will allow commanders to see in a virtual world the effects of new tactics or hardware – a fundamentally new capability. Rather than merely recreating the world, in short, elaborate simulations might someday be powerful enough to change it.

The material on pages 194-7 first appeared in *The Economist* in December 2004.

The cell of a new machine

Is the new Cell chip really as revolutionary as its proponents claim?

ANALOGIES ARE OFTEN DRAWN between the fields of computer science and biology. The information-processing abilities of DNA are a form of natural molecular computing, and computer viruses that leap from machine to machine are examples of artificial, digital biology. At the International Solid-State Circuits Conference in San Francisco in February 2005, a trio of mighty information-technology firms – Sony, Toshiba and IBM – pushed the analogy a little further. They unveiled a much anticipated new computer chip, four years in the making, the very name of which is a biological metaphor: the Cell.

As its name suggests, the Cell chip is designed to be used in large numbers to do things that today's computers, most of which are primitive machines akin to unicellular life-forms, cannot. Each Cell has as its "nucleus" a microprocessor based on IBM's power architecture. This is the family of chips found inside Apple's Power Mac G5 computers and IBM's powerful business machines. The Cell's "cytoplasm" consists of eight "synergistic processing elements". These are independent processors that have a deliberately minimalist design in order, paradoxically, to maximise their performance.

A program running on a Cell consists of small chunks, each of which contains both programming instructions and associated data. These chunks can be assigned by the nucleus to particular synergistic processors inside its own Cell or, if it is deemed faster to do so, sent to another Cell instead. Software chunks running on one Cell can talk to chunks running on other Cells, and all have access to a shared main memory. Since chunks of software are able to roam around looking for the best place to be processed, the performance of a Cell-based machine can be increased by adding more Cells, or by connecting several Cell-based machines together.

This means that programs designed to run on Cell-based architecture should be able to fly along at blistering speeds – and will run ever faster as more Cells are made available. The prototype Cell described in San Francisco runs at 256 gigaflops (a flop – one "floating-point" operation per second – is a measure of how fast a processor can perform the individual operations of digital arithmetic that all computing ultimately boils down to). A speed of 256 gigaflops is around ten times the performance of the chips found in the fastest desktop PCs today; the Cell is thus widely referred to as a "supercomputer on a chip", which is an exaggeration, but not much of one. On the top500.org list of the world's fastest computers, the bottom-ranked machine has a performance of 851 gigaflops.

Hard cell

If you believe the hype, this has other chipmakers – notably Intel, whose Pentium series is the market leader – quaking in their boots. But it is not yet clear that Cell will sweep all before it. The reason is that existing programs have not been designed in the chunky way required if they are to run on Cell-based machines – and rewriting them would be a monumental task.

For the moment, that will not worry Cell's designers because the kinds of things Cell chips are intended to be used for require specially designed software anyway. Cell chips are well suited to processing streams of video and sound, and for modelling the complex three-dimensional worlds of video games, so Cell's debut will be in Sony's next-generation games console, the PlayStation 3. Cell chips will also be ideal for use inside consumer-electronics devices such as digital video-recorders and high-definition televisions. Both Sony and Toshiba plan to use Cell chips in such products. For its part, IBM is talking up the Cell's potential to power supercomputers, the fastest of which, IBM's Blue Gene/L, consists of thousands of special chips that are, in many ways, more primitive versions of Cell. Using Cell chips instead would not, therefore, be a big stretch. And supercomputer programmers, like video-game designers, do not mind learning to program an entirely new machine provided it delivers the performance they are seeking.

If Cell did eventually break out of these specialist applications and into general-purpose computers, Intel would have every right to be paranoid. But Kevin Krewell, the editor of *Microprocessor Report*, an industry journal, sounds a note of caution. The Cell is too power-hungry for handheld devices, and it would need to have its mathematical functions tweaked to be really suitable for use in supercomputers. The Cell is impressive, but, in Mr Krewell's view, "it is no panacea for all those market segments". Similar claims to those now being made for Cell were made in the past about the Sony/Toshiba chip called the Emotion Engine, which lies at the heart of the PlayStation 2. This was also supposed to be suitable for non-gaming uses. Yet the idea went nowhere, and the company set up by Toshiba to promote other uses of the Emotion Engine was closed down.

Extravagant claims were also made about the RISC, POWER and Transmeta architectures, notes Dean McCarron, a chip analyst at Mercury Research in Scottsdale, Arizona. These once-novel methods of chip design have done respectably in specialist applications, but have not dethroned Intel as it was suggested they might when they were launched. Yet both Mr McCarron and Mr Krewell acknowledge that things could be different this time. As Mr McCarron puts it, there are "more ingredients for success present than on previous attempts". Intel was able to see off earlier pretenders to its throne by increasing the performance of its Pentium chips, and by exploiting its economies of scale as market leader. But in this case, the performance gap looks insuperable, and Sony, Toshiba and IBM plan to exploit economies of scale of their own.

Quite how revolutionary the Cell chip will turn out to be, then, remains to be seen. And though it may not be an Intel killer, it could prevent that firm from extending its dominance of the desktop into the living room. Consumer-electronics devices, unlike desktop PCs, do not have to be compatible with existing software. In that sense, the Cell does pose a threat to Intel, which regards the "digital home" as a promising area for future growth. Stand by, therefore, for another round of creative destruction in the field of information technology. And no matter what the Cell does to the broader computer-industry landscape, the virtual game-vistas it will conjure up are certain to look fantastic.

The material on pages 198–200 first appeared in *The Economist* in February 2005.

7
THE DIGITAL HOME

Life in the vault

Companies are fighting to turn your home into an entertainment multiplex

IN JUNE 2004, Intel, the world's largest chipmaker, launched two new lines of chips, code-named Grantsdale and Alderwood, which it called the "most compelling" changes to the way personal computers (PCs) work in "over a decade". From here on, claims Intel, PCs will be "all-in-one hi-fi devices", "entertainment PCs", and "vaults" for digital content.

– Intel's vision is that consumers will start to use their PCs at home to download, store and manage films, songs and games, in order to transmit all this fun stuff wirelessly to TV screens and stereo speakers throughout the house. The kids could then watch *Shrek 2* in the basement, while mum listens to Brahms in the kitchen and dad browses the holiday pictures on the main TV screen in the living room.

As such, Intel's vision is neither new nor overly ambitious. For years, futurists have been peddling notions of digital nirvana in the home. In its wilder forms, this includes fridges that know automatically when to reorder milk via the internet, garage doors that open by themselves as the car approaches, and toilet seats that warm up at just the right moment.

Most of this is guff. Nobody, aside from the self-selected early adopters at trade shows, would consider "upgrading" a garage door every few years for the latest release. "In 20 years, my PC will still not be talking to my fridge," says Jen-Hsun Huang, the chief executive of NVIDIA, the world's largest maker of graphics chips.

On the other hand, says Mr Huang, "the vision of digital content is a much more compelling one than that of home automation." And this is why the new chips may turn out to be as important as Intel claims. They are an opening salvo in a battle between the computer and the consumer-electronics industries over who will dominate the digital household.

Intel, with a virtual monopoly in the chips that power PCs, naturally hopes that PCs will dominate and morph into "media hubs". So does Microsoft, with its near-monopoly on PC operating systems. HP, Gateway, Dell and Apple also want the PC to win, although HP is also big in printers, digital cameras and other consumer gizmos, and Apple has the iPod to fall back on.

On the other side are the giants of consumer electronics. Sony wants future versions of its game consoles, rather than PCs, to play the role of digital "hub". TiVo, a leading maker of digital personal video recorders (PVRs), has hopes for its machines. So do makers of TV set-top boxes.

Extrapolating from history, the PC industry would be the favourite to win, since it has the powerful and rich Microsoft on its side. Microsoft is certainly trying. It has repeatedly relaunched its Windows Media Center, a version of its operating system that looks more like a TV menu and can work via a remote control. Microsoft is also pushing its own next-generation DVD technology, that competes with rival technologies from Japan's Matsushita, NEC, Toshiba and others.

Microsoft's problem is that consumers do not seem keen. Only 32% of American households with internet access polled by Parks Associates, a consumer-technology consultancy, said they were "comfortable" with their PC becoming an entertainment system. Nobody wants to watch the system reboot during a good movie.

This scepticism, however, does not automatically mean that the consumer-electronics industry will win. The one thing that all companies seem to agree on is that households will be connected to the internet via a broadband link that is always on, and that content will be shared wirelessly between rooms within the home. The upshot is that there need not be any single device inside the home that becomes a central media hub. A baby picture could be stored on a PC, on a console, or on a mobile-phone handset. Or it might alternatively be kept on a remote and powerful "server" computer somewhere on the internet. The latter model is how subscribers to Rhapsody, a service provided by RealNetworks, an internet media firm, already listen to music.

The gadget-makers therefore have much to ponder. Art Peck, an analyst at Boston Consulting Group, says that the real money in the digital home will be made by those providing a service or selling advertising. The hardware-makers, he thinks, are therefore fooling themselves by thinking that any device can become a "Trojan horse" to enable them to capture the bounty. It is much more likely that they will all end up as makers of interchangeable commodities for the digital home that the consumer cares little about unless the stuff breaks down.

The material on pages 202–3 first appeared in *The Economist* in July 2004.

Hard disks go home

Hard disks are starting to appear in household devices, from televisions to stereos, adding novel features and making possible new products

GOING ON A LONG TRIP? Desperately afraid of boredom, or silence? Help is at hand. You can now cram 2,000 hours of music – enough for around 120 versions of Wagner's "Ring" cycle – into a device the size of a deck of cards, or squeeze ten hours of video (enough for three or four movies) inside a video player the size of a paperback book. Or perhaps you are stuck at home and want to watch a football game, while simultaneously recording a film and your favourite sitcom on different channels (just to arm yourself against any possibility of boredom in the future). You can do that, too. This is all made possible by a technology normally thought of as part of a personal computer, but now finding its way into a growing range of consumer-electronics devices: the hard-disk drive.

Hard disks have several advantages over other storage media. Unlike the tapes used in video-recorders and camcorders, they do not need to be wound or rewound; disks are "random-access" devices which allow instant jumps from one place to another. Better still, they can also store and fetch more than one stream of data at once, for example to record one TV programme while playing back another.

In some kinds of devices, hard disks also have the edge over solid-state storage media, such as the memory cards used in digital cameras and music players. While hard disks are larger and require more power, they offer far higher capacity – measured in billions of bytes (gigabytes) rather than millions (megabytes) – and at a far lower cost per byte. By and large, hard disks are not used in digital cameras, where small size and long battery life is important, and memory cards are sufficient to store hundreds of images. To some degree the same is true in portable music players as well, but here hard disks can offer more benefit, holding thousands rather than dozens of individual tracks. That is why Apple chose a tiny hard disk for its popular iPod player.

The number of consumer-electronics devices containing hard disks is growing fast, according to figures from InStat/MDR, a market-research company. Around 9m such devices were sold in 2002, and the figure was expected to grow to around 17m in 2003, and reach nearly 90m by 2007 (see Chart 7.1). As well as offering clever new features for consumers, this trend presents a valuable opportunity for hard-disk-makers, which have seen their sales stagnate as the number of PCs sold worldwide has flattened at around 150m units. No wonder they are now eyeing the consumer-electronics market: around 170m TVs, for example, are sold each year.

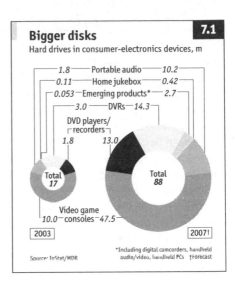

Bigger disks 7.1

Hard drives in consumer-electronics devices, m

- 1.8 — Portable audio — 10.2
- 0.11 — Home jukebox — 0.42
- 0.053 — Emerging products* — 2.7
- 3.0 — DVRs — 14.3
- DVD players/recorders: 1.8 / 13.0
- Video game consoles: 10.0 — 47.5

Total 17 (2003)

Total 88 (2007†)

*Including digital camcorders, handheld audio/video, handheld PCs †Forecast

Source: InStat/MDR

Hard drives are increasingly suitable for use in consumer-electronics devices as they become quieter, cheaper and more robust. Most important of all, they are also getting smaller: some of the biggest potential markets depend on tiny hard drives that appeared on the market only in 2003. In short, the use of hard drives in consumer electronics is still at an early stage. The potential, for both manufacturers and consumers, is vast.

How to save a sitcom

A good example of the use of hard disks in consumer devices is the emerging market for digital video recorders (DVRs). Such devices, pioneered by companies such as TiVo Systems and ReplayTV, have spread most widely thanks to satellite television services such as Dish Network in America and Sky in Britain, both of which incorporate DVR technology into their set-top decoder boxes. DVRs use a hard disk to store video, much like a conventional video recorder, by recording shows at set times. But they may also allow viewers such novelties as pausing and rewinding live television broadcasts (handy for the snack-crazed or those with overactive telephones); recording more than one programme at a time; or recording one programme while playing back another. DVRs can even learn their users' preferences and record programmes accordingly, thus creating the equivalent of a personal TV channel. These are all feats that conventional video recorders cannot match.

Originally, DVRs were built around fairly conventional hard-disk technology. But manufacturers now cater for the DVR market, for example with the creation of a standard "time limiting command", which determines how hard drives handle error checking. Hard drives in PCs are constantly checking and rechecking to make sure they don't lose any data, because a single bit out of place can corrupt an entire document or piece of software. Such fastidious error-checking is not so vital when recording streams of video, where fast and smooth playback is important, and a few lost bits won't affect the image quality. The big harddrive-makers each once had their own approach to time limiting, but they have now agreed on a single standard, which will make it easier for consumer-electronics firms to design new products.

DVRs are still a nascent technology, in use in around 6% of American households at the end of 2004. People who use a DVR almost never want to go back to watching conventional television. Michael Powell, America's former communications regulator, famously described TiVo as "God's machine". But it will be many years before DVRs become ubiquitous. One problem is that consumers often do not understand what they will gain from a DVR until they have used one, so educating the market will take time. DVRs have, however, featured in the plots of sitcoms such as *Sex and the City*, which is a sure sign of their cultural potency.

One thing that might help to spread the word about DVRs is the emergence of handheld video players that use hard drives for storing programmes, such as the Archos AV, which can store up to 80 hours of video using the MPEG4 compression algorithm, and the RCA Lyra. Such devices could provide "TiVo to go", by recording programmes so they can be watched while on the road, and demonstrating how hard disks can transform the experience of watching TV.

A whole new game

While TV presents a huge potential market for hard drives, you are currently more likely to find one inside a different box under the television: a games console. Microsoft's Xbox has a hard drive built in as standard, and a hard-disk attachment can be added to Sony's PlayStation 2. Its next-generation console, the PlayStation 3, will undoubtedly contain a hard drive when it is launched in 2006. Since tens of millions of consoles are sold every year, the emergence of the hard disk as a standard component represents another opportunity for hard-diskmakers.

But since games are supplied on DVD-like disks, why do consoles need a hard drive? With an Xbox connected to the internet via a broadband connection, "you can download new levels and new characters for your games," enthuses Rob Pait of Seagate Technology, a hard-drive-maker. Not everyone is convinced. "If he's doing that, he's an exception," says Danielle Levitas of IDC, a market-research firm. Even publishers working closely with Microsoft on the Xbox have not, she says, figured out how to take advantage of the built-in hard disk: few of the 20m or so Xbox users are actually downloading things to their hard disks. Ms Levitas agrees that consumer-electronics devices will provide a huge new market for hard-disk-makers, but notes that it is taking longer than expected.

A third product category where hard disks are making an impact is in portable music players. Hard disks are ideal for storing music, for while a 60-gigabyte drive can hold around 20 hours of high-quality video, it can hold more music than most people own. The best known example of a disk-based music player is Apple's iPod. But the hard drive faces far more competition in the portable music-player market than it does in set-top boxes or games consoles. Solid-state memory is far more durable than even the most shock-proof hard drive, and consumes less power. And while the cost per byte is much lower for hard disks, the smallest hard disk costs much more than a small (say, 64-megabyte) memory card. That means hard disks cannot compete at the price-sensitive lower end of the music-player market and explains why Apple launched a solid-state version of the iPod, the iPod Shuffle, in 2005.

As a result, memory-based players outsold disk-based ones by 2.8m to 1m in 2002, according to IDC. The company predicts that by 2007, memory-based players will still be ahead, selling 8.2m units, compared with 4.8m disk-based players. And devices that play CDs or MiniDiscs will remain the dominant form of music player for some time, with a combined total of 24m units expected to be sold in 2007, according to IDC's forecasts.

Small is beautiful

The original iPod was built around a particularly small hard disk, the spinning innards of which measure just 1.8 inches in diameter. Most consumer-electronics devices use 2.5-inch or 3.5-inch hard drives, just like laptop and desktop PCs. All of these sizes present an obvious limit: they are too big to fit in a mobile phone, or a small digital camera. But a new technology could carry hard drives into new markets, and help them win a bigger share of existing markets, such as that for music players. That technology is a new generation of 1-inch hard drives, such as Hitachi's Microdrive and a rival product from Cornice, a start-up based in Longmont, Colorado.

Cornice designed its hard drive from scratch, rather than simply scaling down an existing design. The first version of the Cornice drive held only 1.5 gigabytes of data (compared with the 4-gigabyte capacity of Hitachi's 1-inch drive). But it was quickly incorporated into a dozen new products. Some are just smaller versions of existing products, such as RCA's Micro Lyra music player, which is about the size of a small bar of hotel soap. But some are in categories where hard drives have not previously been used such as disk-based camcorders. Apple also used a 1-inch drive to create a smaller version of its iPod music player, the iPod Mini.

Kevin Magenis, the boss of Cornice, says hundreds of companies are designing products around his company's 1-inch drive, from slot machines to a portable karaoke player. By 2008, he claims, Cornice drives will be able to hold 15 gigabytes, expanding the potential market. As well as boosting the capacity of its existing drives, the company also plans to make even smaller ones that can fit inside mobile phones. "That's the killer app for us, but it's a couple of years off," he says.

A hard disk in every pocket? With over 650m mobile phones sold every year, that would open up an enormous new market. No wonder hard-disk makers hope their products will break out of the computer industry. The much bigger world of consumer electronics beckons.

POSTSCRIPT

In 2005, a new generation of even smaller, 0.85-inch hard disks appeared. As the size of hard disks continues to shrink and capacity increases, they are starting to appear in mobile phones. In 2004, Samsung launched a mobile phone with a built-in 1.5 gigabyte hard disk, the SPH-V5400. In 2005, Nokia announced its first handsets with built-in hard disks, which were expected to go on sale by the end of that year.

The material on pages 204–8 first appeared in *The Economist* in December 2003.

A brief history of Wi-Fi

Few people have a kind word to say about telecoms regulators. But the success of Wi-Fi shows what can be achieved when regulators and technologists work together

IT STANDS as perhaps the signal success of the computer industry in the past few years, a rare bright spot in a bubble-battered market: Wi-Fi, the short-range wireless broadband technology. Among geeks, it has inspired a mania unseen since the days of the internet boom. Tens of millions of Wi-Fi devices were sold in 2004, including the majority of laptop computers. Analysts predict that 100m people will be using Wi-Fi by 2006. Homes, offices, colleges and schools around the world have installed Wi-Fi equipment to blanket their premises with wireless access to the internet. Wi-Fi access is available in a growing number of coffee-shops, airports and hotels too. Yet at the turn of the century wireless networking was a niche technology. How did Wi-Fi get started, and become so successful, in the depths of a downturn?

Wi-Fi seems even more remarkable when you look at its provenance: it was, in effect, spawned by an American government agency from an area of radio spectrum widely referred to as "the garbage bands". Technology entrepreneurs generally prefer governments to stay out of their way: funding basic research, perhaps, and then buying finished products when they emerge on the market. But in the case of Wi-Fi, the government seems actively to have guided innovation. "Wi-Fi is a creature of regulation, created more by lawyers than by engineers," asserts Mitchell Lazarus, an expert in telecoms regulation at Fletcher, Heald & Hildreth, a law firm based in Arlington, Virginia. As a lawyer, Mr Lazarus might be expected to say that. But he was also educated as an electrical engineer – and besides, the facts seem to bear him out.

In the beginning

Wi-Fi would certainly not exist without a decision taken in 1985 by the Federal Communications Commission (FCC), America's telecoms regulator, to open several bands of wireless spectrum, allowing them to be used without the need for a government licence. This was an unheard-of move at the time; other than the ham-radio channels, there was very little unlicensed spectrum. But the FCC, prompted by a visionary engineer on its staff, Michael Marcus, took three chunks of spectrum from the industrial, scientific and medical bands and opened them up to communications entrepreneurs.

These so-called "garbage bands", at 900MHz, 2.4GHz and 5.8GHz, were already allocated to equipment that used radio-frequency energy for purposes other than communications: microwave ovens, for example, which use radio waves to heat food. The FCC made them available for communications purposes as well, on the condition that any devices using these bands would have to steer around interference from other equipment. They would do so using "spread spectrum" technology, originally developed for military use, which spreads a radio signal out over a wide range of frequencies, in contrast to the usual approach of transmitting on a single, well-defined frequency. This makes the signal both difficult to intercept and less susceptible to interference.

The 1985 ruling seems visionary in hindsight, but nothing much happened at the time. What ultimately got Wi-Fi moving was the creation of an industry-wide standard. Initially, vendors of wireless equipment for local-area networks (LANs), such as Proxim and Symbol, developed their own kinds of proprietary equipment that operated in the unlicensed bands: equipment from one vendor could not talk to equipment from another. Inspired by the success of Ethernet, a wireline-networking standard, several vendors realised that a common wireless standard made sense too. Buyers would be more likely to adopt the technology if they were not "locked in" to a particular vendor's products.

In 1988, NCR Corporation, which wanted to use the unlicensed spectrum to hook up wireless cash registers, asked Victor Hayes, one of its engineers, to look into getting a standard started. Mr Hayes, along with Bruce Tuch of Bell Labs, approached the Institute of Electrical and Electronics Engineers (IEEE), where a committee called 802.3 had defined the Ethernet standard. A new committee called 802.11 was set up, with Mr Hayes as chairman, and the negotiations began.

The fragmented market meant it took a long time for the various vendors to agree on definitions and draw up a standard acceptable to 75% of the committee members. Finally, in 1997, the committee agreed on a basic specification. It allowed for a data-transfer rate of two megabits per second, using either of two spread-spectrum technologies, frequency hopping or direct-sequence transmission. (The first avoids interference from other signals by jumping between radio frequencies; the second spreads the signal out over a wide band of frequencies.)

The new standard was published in 1997, and engineers immediately began working on prototype equipment to comply with it. Two variants, called 802.11b (which operates in the 2.4GHz band) and 802.11a (which operates in the 5.8GHz band), were ratified in December 1999 and January 2000 respectively. 802.11b was developed primarily by Richard van Nee of Lucent and Mark Webster of Intersil (then Harris Semiconductor).

Companies began building 802.11b-compatible devices. But the specification was so long and complex – it filled 400 pages – that compatibility problems persisted. So in August 1999, six companies – Intersil, 3Com, Nokia, Aironet (since purchased by Cisco), Symbol and Lucent (which has since spun off its components division to form Agere Systems) – got together to create the Wireless Ethernet Compatibility Alliance (WECA).

A rose by any other name ...

The idea was that this body would certify that products from different vendors were truly compatible with each other. But the terms "WECA compatible" or "IEEE802.11b compliant" hardly tripped off the tongue. The new technology needed a consumer-friendly name. Branding consultants suggested a number of names, including "FlankSpeed" and "DragonFly". But the clear winner was "Wi-Fi". It sounded a bit like hi-fi, and consumers were used to the idea that a CD player from one company would work with an amplifier from another. So Wi-Fi it was. (The idea that this stood for "wireless fidelity" was dreamed up later.)

The technology had been standardised; it had a name; now Wi-Fi needed a market champion, and it found one in Apple, a computer-maker renowned for innovation. The company told Lucent that, if it could make an adapter for under $100, Apple would incorporate a Wi-Fi slot into all its laptops. Lucent delivered, and in July 1999 Apple introduced Wi-Fi as an option on its new iBook computers, under the brand name AirPort. "And that completely changed the map for wireless networking," says Greg Raleigh of Airgo, a wireless start-up based in Palo Alto, California. Other computer-makers quickly followed suit. Wi-Fi caught on with consumers just as corporate technology spending dried up in 2001.

Wi-Fi was boosted by the growing popularity of high-speed broadband internet connections in the home; it is the easiest way to enable several computers to share a broadband link. To this day, Wi-Fi's main use is in home networking. As the technology spread, fee-based access points known as "hotspots" also began to spring up in public places

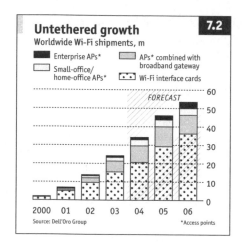

Untethered growth `7.2`

Worldwide Wi-Fi shipments, m

- ■ Enterprise APs*
- □ Small-office/home-office APs*
- ▨ APs* combined with broadband gateway
- ⬚ Wi-Fi interface cards

FORECAST

60
50
40
30
20
10
0

2000 01 02 03 04 05 06

Source: Dell'Oro Group

*Access points

such as coffee-shops, though many hotspot operators have gone bust and the commercial viability of many hotspots is unclear. Meanwhile, the FCC again tweaked its rules to allow for a new variant of Wi-Fi technology, known as 802.11g. It uses a new, more advanced form of spread-spectrum technology called orthogonal frequency-division multiplexing (OFDM) and can achieve speeds of up to 54 megabits per second in the 2.4GHz band.

Where next? Many Wi-Fi enthusiasts believe it will sweep other wireless technologies aside: that hotspots will, for example, undermine the prospects for third-generation (3G) mobile-telephone networks, which are also intended to deliver high-speed data to users on the move. But such speculation is overblown. Wi-Fi is a short-range technology that will never be able to provide the blanket coverage of a mobile network. Worse, subscribe to one network of hotspots (in coffee-shops, say) and you may not be able to use the hotspot in the airport. Ken Denman, the boss of iPass, an internet-access provider based in Redwood Shores, California, insists that things are improving. Roaming and billing agreements will, he says, be sorted out within a couple of years.

By that time, however, the first networks based on a new technology, technically known as 802.16 but named WiMax, should be up and running. As its name suggests, WiMax is positioned as a wide-area version of Wi-Fi. It has a maximum throughput of 70 megabits per second, and a maximum range of 50km, compared with 50m or so for Wi-Fi. Where Wi-Fi offers access in selected places, like phone boxes once did, WiMax could offer blanket coverage, like mobile phones do.

Wi-Fi is also under threat in the home. At the moment it is the dominant home-networking technology: Wi-Fi-capable televisions, CD players and video-recorders and other consumer-electronics devices are already starting to appear. This will make it possible to pipe music, say, around the house without laying any cables. Cordless phones based on Wi-Fi are also in the works. But Wi-Fi may not turn out to be the long-term winner in these applications. It is currently too power-hungry for handheld devices, and even 802.11g cannot reliably support more than one stream of video. And a new standard, technically known as 802.15.3 and named WiMedia, has been specifically designed as a short-range, high-capacity home networking standard for entertainment devices.

Wi-Fi's ultimate significance, then, may be that it provides a glimpse of what will be possible with future wireless technologies. It has also changed the way regulators and technologists think about spectrum policy. The FCC has just proposed that broadcast "whitespace" – the air-waves assigned to television broadcasters but not used for technical rea-sons – should be opened up too. That is not to say that spectrum licensing will be junked in favour of a complete free-for-all over the air-waves. Julius Knapp, the deputy chief of the office of engineering and technology at the FCC, maintains that both the licensed and unlicensed approaches have merit.

Wi-Fi also shows that agreeing on a common standard can create a market. Its example has been taken to heart by the backers of WiMax. Long-range wireless networking gear, like short-range technology before it, has long been dominated by vendors pushing proprietary standards, none of which has been widely adopted. Inspired by Wi-Fi's success, the vendors have now thrown their weight behind WiMax, a common stan-dard with a consumer-friendly name, which they hope will expand the market and boost all their fortunes. Whatever happens to Wi-Fi in future, it has blazed a trail for other technologies to follow.

The material on pages 209–13 first appeared in The Economist in June 2004.

Untangling ultrawideband

Which technology will prevail in the battle to banish the spaghetti behind your TV and computer?

AS ANYONE who has set up a Wi-Fi wireless computer network in their home will attest, there is something extraordinarily liberating about surfing the internet without wires. Perhaps it is because computers, unlike telephones (which went wireless, or at least cordless, a few years ago), always seem to gather a complicated tangle of cables around them. Similarly, another wireless technology, Bluetooth, is starting to do away with the cables between mobile phones and laptops. But other wires stubbornly resist replacement by wireless technology, particularly those that carry video signals – from a DVD player to a television, for example, or from a camcorder to a PC.

Partly, that is because beaming video around reliably requires a higher rate of data transfer than Wi-Fi or Bluetooth can provide. Another problem is that, as wireless transmission speeds increase, so too does power consumption. Wi-Fi-enabled handheld computers, for example, need to be recharged every day, whereas mobile phones, which transfer data much more slowly, can run for a week between charges. The ideal cable-replacement technology, then, would combine blazing speed with low power consumption. And that explains the current interest in an unusual wireless technology called "ultrawideband" (UWB).

UWB has been around for many years in various forms. But it is about to make its first appearance in consumer-electronics products. This ought to be cause for rejoicing, for UWB is a low-power technology that supports data-transfer rates measured in hundreds of megabits per second over short distances (such as between two devices in the same room). UWB thus has the potential to do away with the spaghetti behind computers and home-entertainment systems. It will allow camcorders and digital cameras to beam images directly to televisions or PCs. It could even enable your computer to update your portable music player with your latest downloads automatically as you walk past.

There is just one small problem: the consumer-electronics industry is riven by disagreement, akin to the fight between VHS and Betamax video formats, over which of two versions of UWB to adopt. In one corner is the UWB Forum, which has fewer supporters, but whose technology was expected to appear in consumer devices by the end of 2005. In the other corner is the Multiband OFDM Alliance (MBOA), which has a far more impressive list of backers, but whose products were not expected to reach the market until early 2006. Both sides claim that their version of UWB is superior and will win in the marketplace. But the ironic result is that the great untangler – the technology that was supposed to do away with rats' nests of cables – is itself tangled up in a standards war.

Widespread appeal

The two incarnations of UWB are variations on the same highly unusual technological theme. Unlike conventional radio transmitters, which transmit on a particular frequency and which cannot be picked up if the receiver is slightly mistuned, UWB devices broadcast at very low power over an extremely wide band of frequencies. This has the advantage that UWB signals can be picked up by suitably designed receivers, but resemble background noise to conventional radio receivers, which are listening on one particular frequency. Conventional and UWB radios can therefore coexist. And that is why America's telecoms regulator, the Federal Communications Commission (FCC), ruled in February 2002 that UWB devices could operate across a broad swathe of the radio spectrum, from 3.1GHz to 10.6GHz, without requiring spectrum licences.

This unusual approach makes UWB very different from Wi-Fi and Bluetooth, two other unlicensed radio technologies. Rather than operating (as Wi-Fi and Bluetooth do) in unlicensed "garbage bands", the radio equivalent of unused wasteland, UWB devices operate across frequency bands that are already licensed for various other purposes, including satellite broadcasts, global-positioning systems and telematics. By keeping power levels low, however, UWB devices can coexist with these existing systems – an approach known as "underlay access". Where Wi-Fi exploits the radio equivalent of wasteland, UWB is like being able to build underground. Its novel approach liberates huge amounts of hitherto untapped transmission capacity.

The two sides in the UWB standards war disagree over how best to spread signals out over the radio spectrum, however. The approach favoured by the UWB Forum is called direct-sequence ultrawideband (DS-UWB). A stream of data is combined with a constantly changing pseudo-random code to produce a wideband signal that resembles random background noise. But a receiver armed with the same pseudo-random code can "de-spread" the signal – in effect, plucking it from the background hiss.

UWB chips based on this principle were developed by XtremeSpectrum, a start-up based in Vienna, Virginia. Its UWB assets were then acquired by Freescale, the former chipmaking arm of Motorola, a telecoms-equipment maker. Freescale's first UWB chip, capable of transmitting data at 110 megabits per second, was approved by the FCC in August 2004. Martin Rofheart, the co-founder of XtremeSpectrum and now the head of Freescale's UWB operation, said faster chips, capable of speeds as high as 1 gigabit per second, would be available in 2005.

The rival approach, backed by the MBOA, is called multiband orthogonal frequency-division multiplexing (MB-OFDM). It differs from DS-UWB in several ways. For one thing, rather than spreading an ultrawideband signal right across the allowed UWB frequency range, it divides the range up into 15 bands, each of which is still extremely wide by the standards of conventional radio technology, and constantly hops from one to another. Within each band, the encoding of data is done using a trendy technique called OFDM, which uses elaborate signal-processing techniques to sprinkle information into 128 sub-bands to produce a signal that resembles random noise but can be decoded using a clever-enough receiver.

Eric Broockman of Alereon, a start-up based in Texas that is one of the founders of the MBOA, said prototype chips based on this approach would be available by the end of 2004. The chips were duly announced in October of that year.

So, which technology is better? From a technical standpoint, both have their pros and cons. MB-OFDM is so computationally intensive that it requires ten times as much power as DS-UWB, claims Dr Rofheart, and is therefore less suitable for use in portable devices. And while Freescale's production lines are already up and running, the MB-OFDM camp has yet to produce a prototype – which means, says Dr Rofheart, that they are two years behind, given typical development times for wireless chips.

Inevitably, Mr Broockman disputes all of this. Calculations suggest that the MB-OFDM approach will be perfectly suitable for use in portable devices, he says. Freescale is ahead, Mr Broockman concedes, but by six months, not two years. Besides, says Mr Broockman, all that elaborate signal processing makes MB-OFDM very robust in noisy environments – and UWB devices, by definition, operate in noisy environments, since they have to coexist with existing radio technologies. And the multiband, frequency-hopping approach means the technology can be more easily adapted for use in different parts of the world, where regulators are likely to apply different rules to UWB.

In short, DS-UWB has the advantage of being available sooner, while MB-OFDM is technically more elegant. "This is not a race to write a specification, but to deliver functionality," says Dr Rofheart. "They are ahead with a Model T, and we are pulling up with a Ferrari," retorts Mr Broockman.

All of this has led to deadlock at the body that has been trying to devise a standard for UWB, the so-called 802.15.3 committee at the Institute of Electrical and Electronics Engineers (IEEE). Neither side has been able to garner the necessary 75% support, and things seem likely to stay that way. But there are three kinds of standards, notes Mr Broockman. There are official standards, such as those drawn up by the IEEE; there are industry standards, where big firms team up and agree to adopt a technology (as happened with Bluetooth, for example); and there are de facto standards, decided in the marketplace, as happened with the "Wintel" standard in the PC industry. With deadlock at the IEEE, the MBOA is taking the industry-standard approach, while the UWB Forum is hoping to establish a de facto standard.

And while the UWB Forum has the advantage of having got to market first, the MBOA has far more powerful backers. Its members include chipmaking giants such as Intel and Texas Instruments, consumer-electronics firms including Sony, Matsushita, Philips and Samsung, and other heavyweights including Microsoft, Hewlett-Packard and Nokia. The members of the rival UWB Forum, in contrast, are rather less well known: its most prominent backer is Freescale. That suggests that DS-UWB will have the market to itself for a few months, but will then succumb to the MBOA steamroller. Dr Rofheart, however, claims that many members of the MBOA, including some large consumer-electronics firms, are already testing his firm's chips, and are ready to switch camps.

The dark horse

While the two technologies fight it out, however, there is a third possibility: that a forthcoming form of Wi-Fi, a high-speed technology called 802.11n, might benefit from the confusion and end up stealing some or all of the market for UWB technology. Existing versions of Wi-Fi are already appearing in some consumer-electronics devices. "UWB has a 50–50 shot at the mass market," says Rajeev Chand of

Rutberg & Company, an investment bank in San Francisco. UWB has greater bandwidth and consumes less power, he notes, but Wi-Fi is out now, and engineers are very good at taking an inadequate technology and making it good enough. But both camps of UWB supporters insist that Wi-Fi and UWB will coexist. Wi-Fi, they claim, will be used for piping data around a home network, while UWB will be used to connect devices in the same room.

So it could be a year or two before a clear winner emerges. In the meantime, confusion will reign, and consumers should tread carefully. There is one simple thing that proponents of the rival UWB technologies can do to improve their chances, however. As the successes of Wi-Fi and Bluetooth show, it helps if your technology has a snappy name.

The material on pages 214–18 first appeared in *The Economist* in September 2004.

The meaning of iPod

How Apple's iPod music-player and its imitators are changing the way music is consumed

WHAT IS THE MEANING OF IPOD? When Apple, a computer-maker, launched its pocket-sized music-player in October 2001, there was no shortage of sceptical answers. Critics pointed to its high price – at $399, the iPod cost far more than rival music players – and to the difficulty Apple would have competing in the cut-throat consumer-electronics market. Worse, Apple launched the iPod in the depths of a technology slump. Internet discussion boards buzzed with jokes that its name stood for "idiots price our devices" or "I prefer old-fashioned discs."

Such criticisms were quickly proven wrong. The iPod is now the most popular and fashionable digital music-player on the market, which Apple leads (see Chart 7.3 overleaf). At times, Apple has been unable to meet demand. On the streets and underground trains of New York, San Francisco and London, iPod users (identifiable by the device's characteristic white headphone leads) are ubiquitous. Fashion houses make iPod cases; pop stars wear iPods in their videos. The iPod is a hit.

Its success depends on many factors, but the most important is its vast storage capacity. The first model contained a five gigabyte hard disk, capable of holding over 1,000 songs. The latest models, with 60 gigabyte drives, can hold 15,000. Before the iPod, most digital music players used flash-memory chips to store music, which limited their capacity to a few dozen songs at best. Apple correctly bet that many people would pay more for the far larger capacity of a hard disk. Apple's nifty iTunes software, and the launch of the iTunes Music Store, from which music can be downloaded for $0.99 per track, also boosted the iPod's fortunes.

It is easy to dismiss the iPod as a fad and its fanatical users as members of a gadget-obsessed cult. But the 15m or so iPod users worldwide are an informative minority, because iPod-like devices are the future of portable music. So what iPod users do today, the rest of us will do tomorrow. Their experience shows how digital music-players will transform the consumption of music.

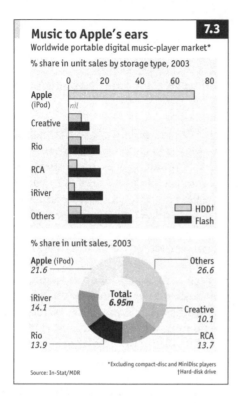

Music to Apple's ears **7.3**
Worldwide portable digital music-player market*
% share in unit sales by storage type, 2003

Apple (iPod) *nil*
Creative
Rio
RCA
iRiver
Others

HDD†
Flash

% share in unit sales, 2003

Apple (iPod) 21.6
iRiver 14.1
Rio 13.9
Total: 6.95m
Others 26.6
Creative 10.1
RCA 13.7

*Excluding compact-disc and MiniDisc players
†Hard-disk drive
Source: In-Stat/MDR

Professor iPod speaks

Few people know more about the behaviour of iPod users than Michael Bull, a specialist in the cultural impact of technology at the University of Sussex in Britain. Having previously studied the impact of the cassette-based Sony Walkman, he is now surveying hundreds of iPod users. Their consumption of music, he says, changes in three main ways.

The first and most important is that the iPod grants them far more control over how and where they listen to their music. Surely, you might ask, an iPod is no different from a cassette or CD-based player, since you can always carry a few tapes or discs with you? But most people, says Dr Bull, find that if none of the music they are carrying with them fits their mood, they prefer not to listen to music at all. The large capacity of a hard-disk-based player does away with this problem. The right music can always be summoned up depending on your mood, the time of day and your activity, says Dr Bull. As a result, iPod users tend to listen to particular music during specific journeys or activities, such as commuting to work or jogging.

By granting them control over their environment – the audible environment, at least – the iPod allows its users to escape into their own little private bubbles. When standing in line at the airport, or waiting for a late train, iPod users feel that not everything, at least, is out of their control. They are also, says Dr Bull, far more selective about answering their mobile phones. That suggests that adding phone functionality to the iPod would be a bad idea, since it would facilitate intrusion.

This does not mean the iPod is inherently anti-social, however. For its second effect is to make music consumption, a traditionally social activity, even more so. You can use your iPod as a jukebox at home, and the ability to carry your music collection with you means you can always play new tracks to your friends. Many iPod users compile special selections of tunes, or playlists, for family listening while in the car. Family members negotiate the contents of the playlist, so that Disney tunes end up juxtaposed with jazz and Justin Timberlake.

That leads to the third of the iPod's effects on music consumption. The ability to mix and match tracks in playlists unconstrained by the limitations of vinyl records or CDs could undermine the notion of the album as a coherent collection of music. Musicians can still make albums if they want to, of course. But with music sold online on a track-by-track basis, albums could suddenly look very old-fashioned, and singles might make a comeback.

Are video iPods next? Strikingly, none of these shifts in usage patterns applies to video. People do not watch movies while walking the dog, make playlists of their favourite movie scenes, or clamour to buy individual scenes online. Portable video-players, which are already starting to become available, undoubtedly have their uses, such as providing entertainment during long journeys. But they seem unlikely to be the kind of industry-changing products that the iPod and its imitators have unexpectedly proven to be.

The material on pages 219–21 first appeared in *The Economist* in June 2004.

Music's brighter future

The internet will eventually be wonderful for music buyers, but it is still a threat to today's dominant record labels

"**D**IRTY POP with wonky beats and sleazy melodies" is how the Sweet Chap, aka Mike Comber, a British musician from Brighton, describes his music. The Sweet Chap has no record deal yet, but he has been taken on by IE Music, a London music-management group that also represents megastar Robbie Williams. To get the Sweet Chap known, in 2003 IE Music did a deal to put his songs on KaZaA, an internet file-sharing program. As a result, 70,000 people sampled the tracks and more than 500 paid for some of his music. IE Music's Ari Millar says that virally spreading music like this is the future.

It may indeed be, and nimble small record labels and artist-management firms will certainly get better results as they find ways to reach more people via the internet. But the question facing the music industry is when that future will arrive. And the issue is most urgent for the four big companies that dominate the production and distribution of music – Universal, Sony/BMG, Warner and EMI (see Chart 7.4). So far they have been slow to embrace the internet, which has seemed to them not an opportunity but their nemesis. Rather than putting their product on file-sharing applications, they are prosecuting free-download users for theft. They have certainly been struggling: sales of recorded music shrank by a fifth between 1999 and 2003.

Today, there is more optimism. In the first half of 2004, global physical unit sales of recorded music rose, albeit by a tiny amount. The industry claims that file-sharing has stabilised thanks to its lawsuits. The number of music files freely available online fell from about 1.1 billion in April 2003 to 800m in June 2004, according to IFPI, a record-industry body. That said, internet piracy is rampant, and physical CD piracy continues to worsen.

But big music's attitude towards the internet has changed, too. Since 2000 the big companies have come a long way towards accepting that the internet and digital technology will define the industry's future. Thanks to Apple and its enormously popular iPod music players and iTunes download service, most music executives now believe that people will pay for legal online music. (Although they have mushroomed, legal online downloads account for less than 5% of industry revenues.) The big companies are trying to work out how they can harness the internet. Consequently, they are having to rethink their traditional business models.

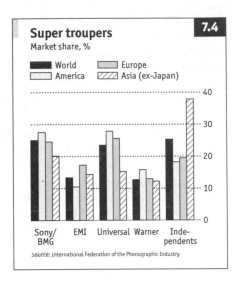

Super troupers 7.4
Market share, %
■ World ▨ Europe
□ America ▨ Asia (ex-Japan)

Sony/ BMG · EMI · Universal · Warner · Independents

Source: International Federation of the Phonographic Industry

In the physical world, the big companies have the advantage of scale. In addition to marketing clout, they own a large back catalogue of music that can be repeatedly reissued. They are also bolstered by music-publishing businesses, which collect royalties on already published songs used in recorded music, live performance, films and advertisements.

Historically, the majors have controlled physical distribution of CDs. Yet that barrier to entry will erode as more music is distributed on the internet and mobile phones. Artists can, in theory, use the internet to bypass record firms, though few have yet done this. The principal reason most have not is that they need marketing and promotion, which the majors also dominate, to reach a wide audience.

The majors have a tight hold on radio, for example, by far the most effective medium for promoting new acts. (Perhaps their lock is too strong: Eliot Spitzer, New York's attorney-general, is investigating whether the companies bribe radio stations to play their music.) Could the internet challenge them on this too? So far, bands have not been launched online. But that could change, and there is already evidence that data derived from the preferences shown on illegal file-sharing networks are being used to help launch acts.

Much will depend on whether the majors choose to address a problem that is just as important as piracy: these days they rarely develop new artists into long-lasting acts, relying instead on short-term hits promoted in mainstream media. That has turned off many potential buyers of new music. In future, using the internet, the industry will be able to appeal directly to customers, bypassing radio, television and big retailers, all of which tend to prefer promoting safe, formulaic acts.

That could give the majors the confidence to back innovative, edgy music. But much smaller independent labels and artist-management firms can do the same, offering them a way to challenge the big firms head on.

Even in the physical world, the big firms are struggling to maintain their traditional market. Supermarkets have become important outlets, but the likes of Wal-Mart stock only a narrow range of CDs, choosing to shift shelf-space away from music in favour of higher-margin DVDs and videogames. That is a symptom of another headache for all music firms: they face ever more intense competition from other kinds of entertainment, especially among the young. In theory, then, digital technology offers the majors an escape hatch. With infinite space and virtually free distribution online, every track ever recorded can be instantly available to music fans. Of course, smaller firms will be able to do the same thing.

Where did all the music go?

According to an internal study done by one of the majors, between two-thirds and three-quarters of the drop in sales in America had nothing to do with internet piracy. No one knows how much weight to assign to each of the other explanations: rising physical CD piracy, shrinking retail space, competition from other media, and the quality of the music itself. But creativity doubtless plays an important part.

Judging the overall quality of the music being sold by the four major record labels is, of course, subjective. But there are some objective measures. A successful touring career of live performances is one indication that a singer or band has lasting talent. Another is how many albums an artist puts out. Many recent singers have toured less and have often faded quickly from sight.

Music bosses agree that the majors have a creative problem. Alain Levy, chairman and chief executive of EMI Music, told *Billboard* magazine in 2004 that too many recent acts have been one-hit wonders and that the industry is not developing durable artists. The days of watching a band develop slowly over time with live performances are over, says Tom Calderone, executive vice-president of music and talent for MTV, Viacom's music channel. Even Wall Street analysts are questioning quality. If CD sales have shrunk, one reason could be that people are less excited by the industry's product. A poll by *Rolling Stone* magazine found that fans, at least, believe that relatively few "great" albums have been produced recently (see Chart 7.5).

Big firms have always relied on small, independent music firms for much of their research and development. Experimental indies signed Bob Marley, U2, Pink Floyd, Janet Jackson, Elvis Presley and many other hit acts. Major record labels such as CBS Records, to be sure, have signed huge bands. But Osman Eralp, an economist who advises IMPALA, a trade association for independent music companies in Europe, estimates that over 65% of the majors' sales of catalogue albums (music that is at least 18 months old) comes from artists originally signed by independents.

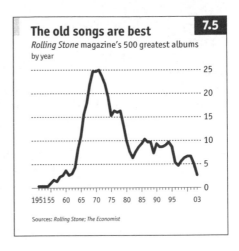

The old songs are best 7.5

Rolling Stone magazine's 500 greatest albums by year

Sources: *Rolling Stone; The Economist*

In the past, an important part of the majors' R&D strategy was to buy up the independent firms themselves. But after years of falling sales and cost-cutting, the majors have little appetite for acquisitions, and now rely more on their own efforts.

What Mr Levy calls music's "disease" – short-term acts – is not solely a matter of poor taste on the part of the big firms. Being on the stockmarket or part of another listed company makes it hard to wait patiently for the next Michael Jackson to be discovered or for a slow-burning act to reach its third or fourth breakthrough album. The majors also complain that the radio business is unwilling to play unusual new music for fear of annoying listeners and advertisers. And while TV loves shows like *Pop Idol* for drawing millions of viewers, such programmes also devalue music by showing that it can be manufactured. Technology has made it easy for music firms to pick people who look good and adjust the sound they make into something acceptable, though also ephemeral.

The majors could argue that they can happily carry on creating overnight hits; so long as they sell well today, why should it matter if they do not last? But most such music is aimed at teenagers, the very age group most likely to download without paying. And back-catalogue albums make a great deal of money. The boss of one major label estimates that, while catalogue accounts for half of revenues, it brings in three-quarters of his profits. If the industry stops building catalogue by relying too much on one-hit wonders, it is storing up a big problem for the future.

A new duet

There are signs that the majors are addressing the issue. Universal Music and Warner Music are starting up units to help independent labels with new artists, both promising initiatives that show that they are willing to experiment. Thanks to the majors' efforts in the last few years, their music has already improved, says Andy Taylor, executive chairman of Sanctuary Group, an independent, pointing to acts such as the Black Eyed Peas (Universal), Modest Mouse (Sony), Murphy Lee (Universal) and Joss Stone (EMI).

And yet even if they can shore up their position in recorded music, the big firms may find themselves sitting on the sidelines. For only their bit of the music business has been shrinking: live touring and sponsorship are big earners and are in fine shape. Between October 2003 and October 2004, according to a manager who oversees the career of one of the world's foremost divas, his star earned roughly $20m from sponsorship, $15m from touring, $15m from films, $3m from merchandise and $9m from CD sales. Her contract means that her record label will share only in the $9m.

In 2002 Robbie Williams signed a new kind of deal with EMI in which he gave it a share of money from touring, sponsorship and DVD sales as well as from CDs, in return for big cash payments. Other record firms are trying to make similar deals with artists. That will be difficult, says John Rose, former head of strategy at EMI and currently a partner at the Boston Consulting Group in New York, because many artists, and their managers, see record companies less as creative and business partners than as firms out to profit from them.

Artists' managers will resist attempts to move in on other revenue streams. Peter Mensch, the New York-based manager of the Red Hot Chili Peppers, Shania Twain and Metallica, says "we will do everything and anything in our power to stop the majors from grabbing any share of non-recorded income from our bands." Mr Mensch says that one way to fight back would be to start his own record company.

Independent labels are also gunning for the big firms. For one thing, they are fighting to stop further consolidation among the majors because that would make it even harder for the independents themselves to compete for shelf space and airplay. IMPALA has mounted a legal challenge against the European Commission's decision allowing

Sony and BMG to merge in 2004. But the small firms are also optimistic that they can grow at the expense of their big rivals. The majors are cutting back in smaller markets and dropping artists who lack the potential to sell in lots of countries. That leaves a space for the indies. For example, Warner Music Group was readying itself for an initial public offering in 2005 and, as part of cutting costs in Belgium, it dropped artists in 2004. Among them was Novastar, whose manager says the group's latest album has so far sold 56,000 copies in Belgium and the Netherlands.

The more the majors scale back, the more the market opens up. People who have left the big firms are starting up new ventures. In January 2005, Emmanuel de Buretel, previously a senior manager at EMI, launched an independent record label called "Because", with help from Lazard, an investment bank. Tim Renner, formerly chairman of Universal Music in Germany, is setting up a music internet service, a radio station in Germany and possibly a new record label.

In the material world

Meanwhile, the majors are trying to plot their move to digital. Making the transition will be tricky. Bricks-and-mortar music retailers need to be kept happy despite the fact that they know that online music services threaten to make them obsolete. It is still unclear what a successful business model for selling music online will look like. People are buying many more single tracks than albums so far. If that persists, it should encourage albums of more consistent quality, since record companies stand to make more money when people spend $12 on a single artist than if they allocate $2 to each of six bands. Or it could mean that the concept of the album will fade.

Online pricing is unstable too. It is likely that download prices will vary in future far more than they do now. Apple forced the industry to accept a fixed fee per download of 99 cents, but the majors will push for variable, and probably higher, prices. Online prices will have an impact on prices in the physical world, which are already gradually falling in most markets. But the result of all these variables might be structurally lower profits.

Edgar Bronfman junior, chairman and chief executive officer of Warner Music Group, expects that paid-for digital-music services via the internet and mobile phones will start to have a measurable impact on music firms' bottom lines as soon as 2006. The new distribution system will connect music firms directly with customers for the first time. It will also shift the balance of power between the industry and giant retailers. Wal-Mart, for instance, currently sells one-fifth of retail CDs in America, but recorded music is only a tiny proportion of its total sales.

The best distribution of all will come when, as many expect, the iPod or some other music device becomes one with the mobile phone. Music fans can already hold their phones up to the sound from a radio, identify a song and later buy the CD. At $3.5 billion in annual sales, the mobile ringtone market has grown to one-tenth the size of the recorded music business.

But can paid-for services compete with free ones? The paying services need to put more catalogue online if they want to match the file-sharing networks with their massive music libraries. And it is still unclear how much "digital-rights management" – technology that restricts how a music download can be used – people will tolerate. Another key issue is interoperability: whether the various new devices for playing digital music will work with other online stores. Apple's iPods, for instance, work with iTunes, but not with Sony Connect or Microsoft's MSN Music Store. Too many restrictions on the paid-for services may entrench file-sharing.

Out of the more than 100 online music sites that exist now, a handful of big players may come to dominate, but there will be specialist providers too, says Ted Cohen, head of digital development and distribution at EMI. iTunes is like the corner store where you buy milk and ice cream, he says, but a customer does not spend much time there. Real Networks' Rhapsody, on the other hand, charges a monthly subscription in return for unlimited streaming music and gives descriptions that lead people to new artists. Recommendation services like these, as well as people sharing playlists, will eventually make the internet a powerful way to market music as well as to distribute it.

Jiving with the enemy

In September 2004, according to comScore Media Metrix, 10m American internet users visited four paid online-music services. The same month another 20m visited file-sharing networks. The majors watch what is being downloaded on these networks, although they do not like to talk about it for fear of undermining their legal campaign.

Online music might truly take off if the majors were to make a truce with the file-sharing networks. The gulf between the two worlds has narrowed now that the industry sells its product online and allows customers to share music using digital-rights management. As for the file-sharing networks, "the other side is more willing to talk and less adversarial," says an executive at one of the majors in Los Angeles.

Music industry executives say that Shawn Fanning, founder of Napster, the first file-sharing network, is working out how to attach prices to tracks downloaded from such services, with a new venture called "Snocap". Mr Fanning tried to make the original Napster legal back in 2001, but the music industry decided instead to sue it out of existence. Snocap has now been licensed by EMI, Sony/BMG, Universal and over 500 independent labels. Sam Yagan, boss of eDonkey, currently the most popular file-sharing network, says he had meetings with three of the four major labels in summer 2003 about how his network could start selling their music alongside free content. As IE Music's experiment shows, that is not an impossible dream. Music executives may not have the confidence yet to make a deal with their arch-enemies. But eventually they have to get bolder. It seems clear that the only way for the majors to stay on top of the music industry into the next decade is to take more risks – both technological and creative – than they have done for a long time.

The material on pages 222–9 first appeared in *The Economist* in October 2004.

Televisions go flat

TVs based on bulky cathode-ray tubes are giving way to flat-panel models. How will the market evolve?

TELEVISIONS, it seems, can never be too wide or too thin – and increasingly, they are wide and thin at the same time, thanks to the growing popularity of flat-panel televisions based on plasma and liquid-crystal display (LCD) technology. Flat-panel TVs are stylish, do not take up much room, and do justice to the crystal-clear images produced by DVD players, digital-cable boxes and games consoles. Sales of LCD TVs in particular are expected to account for an ever larger portion of the market (see Chart 7.6) as consumers embrace these new technologies at the expense of bulky models based on old-fashioned cathode-ray tubes (CRTs). LCD-based models are expected to account for 18% of televisions sold in 2008, up from just 2.2% in 2003, according to iSuppli, a market-research firm.

LCD TVs are the latest example of a technology from the computer industry causing a stir in consumer electronics. For years, anyone who wanted to buy a flat-panel television had to buy a plasma screen, a large and expensive (a 42-inch model costs around $3,500) option. LCD technology, already used in flat-panel computer monitors and laptop displays, makes possible smaller, more affordable flat-panel TVs.

The prospect of a much bigger market has prompted new entrants, including PC-makers such as Dell and HP, and established consumer-electronics firms, such as Motorola and Westinghouse (both of which stopped making TVs decades ago) to start selling televisions alongside the established television-set manufacturers. For PC-makers, which already sell flat-panel monitors, diversifying into TVs is no big leap. For consumer-electronics firms, the appeal of flat-panel TVs is that they offer much higher margins than conventional televisions. During the late-2003 holiday season, makers of flat-panel TVs, both LCD and plasma, succeeded in creating a tremendous buzz around their products, says Riddhi Patel, an analyst at iSuppli.

But it did not translate into sales to the extent that the manufacturers had hoped. Although more people are now aware of flat-panel TVs, many are still deterred by their high prices. The expense is difficult to justify, particularly since a 30-inch LCD television can cost up to four times as much as a comparable CRT-based model, with no real difference in picture quality.

Flat-panel TV-makers have since, says Ms Patel, begun to cut their prices. For one thing, they were sitting on a lot of unsold inventory: the panel-makers made too many panels, the TV-makers built too many TVs, and the retailers ordered more than they could sell.

Prices are also expected to fall as production capacity is

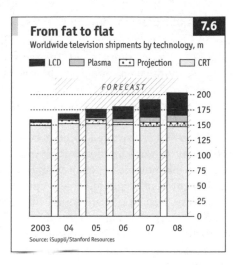

From fat to flat   7.6

Worldwide television shipments by technology, m

■ LCD   ☐ Plasma   ⊡ Projection   ☐ CRT

FORECAST

2003   04   05   06   07   08

Source: iSuppli/Stanford Resources

stepped up. Sharp opened a new "sixth generation" LCD factory in January 2004. In May, Matsushita, the Japanese firm behind the Panasonic brand, announced that it would build the world's biggest plasma-display factory. And in July, Sony and Samsung announced that their joint-venture, a "seventh-generation" LCD factory at Tangjung in South Korea, would start operating in 2005. There was concern that 2004's record investment in LCD plants could lead to overcapacity in 2005. For consumers, however, this is all good news: a glut means lower prices.

The prospect of sharp price declines over the next few years means the flat-panel TV market is on the cusp of change. At the moment, LCD is more expensive than plasma on a per-inch basis: a 30-inch LCD TV costs around the same as a 40-inch plasma model. The vast majority of LCD TVs sold are currently 20 inches or smaller; larger sizes cannot yet compete with plasma on price. So plasma has the upper hand at larger sizes for the time being, while LCDs dominate at the low end.

For anyone looking to buy a flat-panel TV, this makes the choice relatively simple: if you want anything smaller than a 30-inch screen, you have to choose LCD; and if you are thinking of buying bigger, plasma offers better value. (Above 55 inches, TVs based on rear-projection are proving popular, having benefited from the buzz surrounding flat-panel displays.)

Watch out plasma, here comes LCD

As the new LCD plants start running, however, LCD TVs will increasingly be able to compete with plasma at sizes as large as 45 inches. The new seventh-generation LCD plants will crank out screens on glass sheets measuring 1.9 by 2.2 metres, big enough for twelve 32-inch or eight 40-inch panels. LCD could thus push plasma upmarket, unless makers of plasma TVs drop their prices too.

The result is expected to be a fierce battle around the 42-inch mark. This may prompt buyers to look more closely at the relative merits of the two technologies, each of which has its pros and cons. Plasma offers higher contrast, which means deeper blacks. But although the longevity of plasma panels has improved in recent years, from 10,000 hours to 30,000 hours, LCD panels have a lifetime of 60,000 hours. LCD TVs also have the advantage that they can be used as computer monitors. But their response is slower than plasma, so they are less suitable for watching sports.

As production capacity increases and prices fall – they plunged by 40% during 2004 – flat-panel televisions of one variety or another seem destined to become far more widespread. The screens may be flat, but their market prospects are anything but.

The material on pages 230–2 first appeared in *The Economist* in September 2004.

PART 3

SEARCHING FOR THE NEXT BIG THING

Part 3 evaluates the candidates for the next great technological revolution. First is biotechnology, which could make possible personalised medicines and new industrial processes based on genetic engineering, and which has yet to deliver on its true promise, according to our survey, "Climbing the helical staircase". The second candidate is energy technology. A collection of articles examines promising energy technologies including fuel cells, lithium-ion batteries, smart power grids, hybrid cars and green buildings. Third is nanotechnology, the subject of another survey, "Small wonders", which argues that both the hype and the paranoia surrounding this new field are overdone, but that nanotechnology should be welcomed nonetheless. Lastly, two articles consider two technologies, robotics and artificial intelligence, that were touted as the "next big thing" in the past. Now considered failures, they may in fact have been more successful than they appear.

8

CLIMBING THE HELICAL STAIRCASE

Climbing the helical staircase

Biotechnology has its troubles, but in the long term it may change the world

"IT HAS NOT ESCAPED OUR NOTICE that the specific pairing we have postulated immediately suggests a possible copying mechanism for the genetic material." With these ironic words, James Watson and Francis Crick began a biological revolution. Their paper on the structure of DNA, published in *Nature* in April 1953, described the now-famous double helix. It showed that the strands of the helix complement each other. It inferred, correctly, that either strand of the helix could thus act as a template for the other, allowing the molecule to replicate itself. And it suggested that because the four types of nucleotide sub-unit of which each strand is composed can be arranged in any order, a single strand could act as a message tape telling a cell which proteins to make, and therefore what job to do.

As another Francis pointed out four centuries ago, knowledge is power. Sir Francis Bacon's philosophy of turning scientific knowledge to practical advantage eventually delivered the wealth of the industrial revolution. That technology was based mainly on the physical sciences. Now the discoveries made by Dr Watson, Dr Crick and their numerous colleagues, successors, collaborators and rivals are starting to be commercialised as well. Biotechnology is beckoning.

It promises much: more and better drugs; medical treatment tailored to the individual patient's biological make-up; new crops; new industrial processes; even, whisper it gently, new humans. A few of those promises have been delivered already. Many have not. Some may never be. Some may raise too many objections.

But the field is still in its infancy, and commercialising the edge of scientific research is a hazardous business. False starts have been more frequent than successes. The businessmen-scientists who are biotechnology's entrepreneurs often seem driven by motives more complex than a mere desire to make money, especially when they are trying to find treatments for disease. And, at the moment, there is virtually no money for new ventures, leading sceptics to question whether the field has a future at all. But that is to confuse short-term problems with long-term potential. This section will endeavour to cover both, though with greater emphasis on the potential than on the problems. Still, the problems are real and should not be ignored.

Cornucopia or white elephant?

Biotechnology has been through funding crises before, but almost everyone seems to agree that its 2002–03 crisis was the worst yet. Capital to finance new ventures pretty well dried up. Stockmarket flotations stopped, and the share prices of publicly quoted companies declined so much that many firms were worth little more (and sometimes less) than the cash they had in the bank.

It is true that all shares, and shares in high-technology companies in particular, did badly. But it is odd that the biotech sector was punished so severely. For those firms that aspire to be drug companies, the risks have always been high. As John Wilkerson of Galen Associates, a health-care investment company in New York, puts it, "A biotech company is a pharma company without sales." Only a tiny fraction of potential drugs make it through the hazardous process of clinical trials and regulatory approval. But that has always been the case. And the rewards can be high, too. Demand for the drugs that do make it is pretty much guaranteed. Medical need is not tied to the economic cycle. Share prices may have got silly at the end of the 20th century, but biotechnology is not dotcommery. Nothing fundamental has changed.

One possible explanation, offered by Stelios Papadopoulos, vice-chairman of SG Cowen, a New York-based arm of the Société Générale bank, is that in biotechnology fundamentals do not really count. Mr Papadopoulos points out that the fund managers who drove the biotech boom of the late 1990s were themselves driven by bonuses that depended less on how their funds did in absolute terms than on how they did relative to each other. To get a big bonus, managers had to beat the average.

In those circumstances, buying volatile, speculative shares looked like a one-way bet. If they went up, so did your pay packet. If they went down, you were little worse off than if you had bought something safe. And few shares were more speculative than those of a biotechnology company with a handful of wannabe drugs that might fail in clinical trials or be turned down by the regulators. Nevertheless, the fund managers' demand for such "high-beta" stocks pushed up the market for new issues, turning rising prices into a self-fulfilling prophecy. The effect was felt all the way along the chain, from the venture-capital funds to the "angels" – the rich individuals, often themselves successful biotech

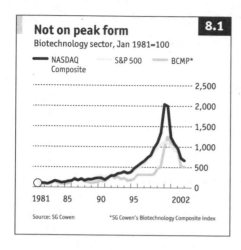

Not on peak form **8.1**

Biotechnology sector, Jan 1981=100

NASDAQ Composite S&P 500 BCMP*

2,500
2,000
1,500
1,000
500
0

1981 85 90 95 2002

Source: SG Cowen *SG Cowen's Biotechnology Composite index

entrepreneurs, who put up seed money to convert promising ideas into business plans.

In a falling market, the incentives for fund managers are different. Safe, "low-beta" stocks are seen as the way to stem losses and outperform colleagues. That creates feedback in the opposite direction, making for another self-fulfilling prophecy. In the absence of stockmarket flotations, venture capitalists are left without an "exit strategy" other than a trade sale to a big pharmaceutical company, which is unlikely to be financially attractive. That means they will stop looking for any new businesses to back and start reducing the numbers already on their books.

For the bold investor, though, such conditions present a buying opportunity. Forward Ventures, a venture-capital firm based in San Diego, believes that firms which have a clear idea of the products they intend to make, and have ways of generating revenue on the slow road to developing them, are worth gambling on. Those that just want to do biological research in the hope that something marketable will turn up are not. Conversely, those that have a range of potential products based on the same technological platform are particularly attractive. Eventually, Forward's partners reckon, someone will come up with a new blockbuster drug, investors will wish they had got a piece of the action and the whole cycle will start all over again.

The old red, white and green

Leaving better medicine to one side, biotechnology has other things to offer too. Many experts in the field categorise its divisions by colour. Red is medical, green is agricultural and white is "industrial" – a broad and increasingly important category that includes making advanced enzymes with a wide variety of uses, and will soon embrace the biotechnological manufacture of plastics and fuel.

Green biotech, too, has its problems. In parts of Europe, in particular, it is beleaguered by militant environmentalists and doubtful consumers.

White has so far remained too invisible to the general public to have stirred up any trouble. That may soon change. Several chemical firms plan to market biotech plastics and artificial fibres on the back of their biodegradability and the fact that they are not made from oil, thus emphasising their environmental friendliness. That should help the makers gain the moral high ground on which producers of genetically modified (GM) crops missed out.

There is one other class of biotechnology. In one sense it is a subset of red, but it is more than that. This is the (as yet hypothetical) biotechnology of manipulating people. Human cloning and genetic engineering are the techniques that dare not speak their names. They are already under assault even though neither yet exists.

This section will examine the practicality and ethics of manipulating people, but first it will look at the grubbier business of products and money, and how to make both of them. On the medical side, most of the innovation has been done by small firms. Big pharmaceutical companies, which increasingly tend to buy, rather than generate, novelty, have been getting less of a look-in. But in the fields of green and white biotech, big firms are often the innovators too.

Perhaps the biggest question is this: will biotechnology remain a niche business, or will it become ubiquitous – as widespread (yet invisible) as the products of the chemical industry are today?

In the 1870s, the science of chemistry was in much the same position as biology is now. It had recently acquired a coherent theoretical framework, the periodic table, which may not have answered all the questions but at least suggested which of them it was sensible to ask. The equivalent for biology is genomics, because a creature's genome is, in a loose way, its own private periodic table of possibility. The industrial chemistry of the 1870s was likewise similar to today's biotechnology. Then, chemists were applying their new, systematic knowledge to a limited range of applications such as dyes and explosives. Existing biotechnology is similarly limited in range. Nowadays, though, it is hard to reach out and touch something that industrial chemistry has not touched first. In a hundred years or so, will the same be true of biotechnology?

Cluster analysis

Like birds of a feather, biotech firms flock together

MARKETING DEPARTMENTS like catchy names. Whether "Biotech Beach" will trip off the tongue as easily as "Silicon Valley" remains to be seen, though the promoters of San Diego certainly hope so. According to a report published in 2001 by the Brookings Institute, a think-tank based in Washington, DC, 94 biotechnology companies were then located in that city. It wants to beat its northerly neighbour, Biotech Bay (ie, San Francisco and its neighbours, which between them have 152), in the competition to become the principal "cluster" of the new industry. That is, if that honour does not fall to Massachusetts, where 141 firms are found clinging to the skirts of MIT and Harvard.

All this sounds faintly familiar, until the realisation dawns: Biotech Bay is simply Silicon Valley by another name. Massachusetts is route 128. The "me-too" clusterettes in Europe – around Cambridge, England, for example, or Uppsala, Sweden – are also in the same places as concentrations of electronics and software companies. The same is true of Israel, which likes to regard itself as one big biotech cluster. But San Diego is the exception to the rule. The local naval base has brought some "hard-edged" technology to the area, but it is hardly a hotbed of IT.

It does, however, have three world-class biological research institutes within walking distance of one another (if anybody did any walking around there): the University of California, San Diego, with its medical school, the Scripps Clinic and the Salk Institute. Most of the biotech ventures are similarly crowded together, some in roads with hopeful names such as Sequence Drive. In other words, this is the most clustered cluster around.

The big pharmaceutical companies are also starting to open laboratories here. Johnson & Johnson, Merck, Novartis and Pfizer have all moved in. The cross-fertilisation of ideas and staff-poaching that this encourages give the place a family atmosphere, helped by the fact that many firms can trace their roots back, one way or another, to just two companies: Hybritech, a diagnostics specialist which was the first biotech success in San Diego, and IDEC Pharmaceuticals, which has been making cancer-fighting antibodies since the mid-1980s. If there is an authentic biotechnology cluster anywhere in the world, this is it. And yes, you can see the beach from some of the labs.

A voyage of discovery

Biotechnology may yet renew the pharmaceuticals industry

IN THE BEGINNING (ie, the late 1970s), doing red biotechnology was easy. You picked a protein that you knew would work as a drug, ideally because it was being used as one already. You copied its DNA into a bacterium or a cell from the ovary of a Chinese hamster. You bred those cells by the billion in big, stainless-steel fermenting vessels known as bioreactors. You extracted the protein. Then you watched the profits roll in. Of course it wasn't really quite as easy as that – but if you could get the manufacturing process right, the chance of making something that worked and would satisfy the regulators was pretty good.

That golden age will never return. Now biotech companies have to find their proteins before they can start making them. Genomics, which involves working out the complete sequence (ie, the order of the nucleotide sub-units) of the DNA in an organism, can help them do that.

Proteins are chains of sub-units called amino acids. The order of the nucleotides in a gene, read off in groups of three, describes the order of the amino acids in a protein. Read the genes and you know what the proteins should look like. That sounds like an attractive short cut, and, as a consequence, genomics caught the attention of both the public and the stockmarkets in the dying years of the 20th century. Several firms thought they could build businesses by collecting and selling genomic information, and two rival groups, the publicly funded International Human Genome Sequencing Consortium and a private company, Celera Genomics, raced to produce versions of the genome of most interest to humanity: its own. Two draft sequences were duly published in 2001, though work on polishing them continues. And lots of new genes have been discovered, with each implying the existence of at least one new protein that might, possibly, have some therapeutic value.

But genomics tells you only so much about a protein. It needs to be backed up with other sorts of -omics, too. There is, for example, proteomics (cataloguing and analysing all the kinds of protein molecule actually produced in an organism). There is transcriptomics (logging the intermediary molecules, known as messenger RNAS, that carry information from the DNA in the nucleus to the rest of the cell). There is glycomics (doing the same for carbohydrate molecules, which often affect the way a protein works) and metabolomics (studying the small molecules that are processed by proteins). There is even bibliomics, which mines the published scientific literature for unexpected connections between all of the above. But, as Sydney Brenner, who won the 2002 Nobel prize for medicine, once wryly observed, in biotechnology the one -omics that really counts is economics. By that yardstick, there is still a long way to go.

Ome on the range

Companies that had placed their faith in genomics are now caught in a dilemma. Investors have decided, probably correctly, that merely collecting -omics information (gen- or otherwise) and selling it to drug companies is not a big enough business to be worth doing in its own right. They would rather put their money behind firms that are trying to develop drugs. The information companies are therefore trying to reinvent themselves. However – and this is where the dilemma comes in – if new drugs are to be discovered, exploiting -omic information is one of the most likely routes to success.

Some companies have understood this since the beginning. Incyte, founded in 1991, and Human Genome Sciences (HGS), set up in 1992, both began by using transcriptomics to see which genes are more or less active than normal in particular diseases. But HGS always saw itself as a drug company, whereas Incyte was until recently more of an information company that would sell its discoveries to others. As a result, HGS now has ten candidate drugs in the pipeline, whereas Incyte has none.

Starting from scratch, it takes a long time to become a real drug company (ie, one with drugs on the market). Millennium Pharmaceuticals, another firm dating from the early 1990s, whose business model was half-way between Incyte's and HGS's, has cut the Gordian knot by buying in drugs developed by others and scaling down its in-house discovery programme.

That should work from a business point of view. So should the decision by Applera, the parent of Celera, to pull back from its grandiose plans to follow up the genome by completing the human proteome. The protein-analysing facility in Framingham, Massachusetts, where this was to be done, is now being used as a test bed for Applera's lucrative scientific-instrument business. The question is whether with hindsight such decisions will be viewed as prudent adjustments to reality or lamentable failures of nerve. And that depends on whether the programmes these firms are now putting on the back burner could eventually have been turned into profitable technological platforms for producing multiple products.

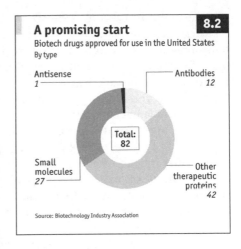

A promising start   8.2
Biotech drugs approved for use in the United States
By type

Antisense
1

Antibodies
12

Total:
82

Small molecules
27

Other therapeutic proteins
42

Source: Biotechnology Industry Association

Not everyone has lost faith in -omics, though. The quest for the proteome has been picked up by Myriad, of Salt Lake City (an early biotech success, which owns the rights to tests for genetic susceptibility to breast cancer). It has formed a collaborative venture with Hitachi, a Japanese electronics firm, and Oracle, an American database company, to identify all the human proteins, and also to work out the interactions between them by expressing their genes in yeast cells and seeing which proteins like to react with one another.

That protein-interaction web will be scientifically invaluable. How much it will profit Myriad remains to be seen. Other proteomics companies, such as Oxford GlycoSciences (which was taken over by Celltech in April 2003), are more interested in comparative proteomics, which involves looking for differences between the proteins in healthy and diseased tissue. The idea is to identify proteins that might make good targets for drugs.

Even genomics still has followers, though the survivors have their eyes firmly on connecting genes to diseases, thus creating drug-discovery platforms. The buzzwords are "SNP" (pronounced "snip") and "haplotype". SNP stands for single-nucleotide polymorphism – a point on a DNA strand where nucleotides can vary from person to person. Groups of SNPs often hang around together, acting as markers for entire blocks of DNA. The combination of these blocks that an individual carries is known as his haplotype, and certain haplotypes seem to be associated with particular diseases. If there is a SNP in a gene, it may cause the protein described by that gene to be abnormal, leading to disease. If it is in the non-coding "junk" DNA that forms about 96% of the human genome, it may still be useful as a marker to follow haplotype blocks around.

Several firms are running SNP/haplotype-based drug-discovery projects. Perhaps the best-known of them is deCODE, an Icelandic company that has recruited much of the country's population for its research. Icelandic genealogical records are so good that the family relationships of most living Icelanders with each other are known. That, plus the country's excellent medical records and the willingness of many people to donate their DNA to the cause, has allowed the firm to follow haplotype blocks down the generations, matching them to the diseases people have reported to their doctors. A search within the relevant block may then reveal a gene associated with the disease.

DeCODE is trying to track down the genetic underpinnings of more than 50 diseases. So far it has found the general location of genes associated with 25 of them, and pinpointed genes for seven, including schizophrenia and strokes. Its method is based on traditional genetics, involving the study of only those blocks of DNA that these techniques suggest are shared.

Perlegen, based in Mountain View, California, and Sequenom, based in San Diego, cast their nets wider by going for all the SNPs, albeit in far fewer individuals than deCODE works with. This is done by "resequencing", in other words studying people's genomes only at the sites such as SNPs where variation is known to occur, and assuming that the rest of the DNA will be the same in everyone.

Perlegen uses sets of special "gene chips" that have short stretches of "probe" DNA containing the complement of every possible SNP dotted over their surfaces. If a SNP is present in a sample, it will stick to the appropriate probe. Sequencing an individual this way costs $2m a pop, and Perlegen has thought it worthwhile to use $100m of its start-up capital recording the genomes of 50 people. Sequenom, a longer-established firm, identifies SNPs by their different molecular weights, in a machine called a mass spectrometer.

Genaissance, another haplotype company, is taking a different tack. Instead of trying to connect genes with diseases, it is connecting them directly with existing drugs, by looking at the way people with different haplotypes respond to different treatments for what appear to be the same symptoms. Its flagship project is studying statins, drugs intended to regulate the level of cholesterol in the blood (a $13 billion market in America alone). Different patients respond differently to each of the four drugs in the trial, and Genaissance is beginning to unravel the reasons for this – or at least to be able to predict from an individual's haplotype which of the four will work best.

This sort of work is obviously valuable from a patient's point of view. Indeed, the hope is that it will one day lead to "personalised medicine", which will identify an individual's disease risks well before a disease appears, and know in advance which drugs to prescribe. Drug companies may feel more equivocal, since even misprescribed drugs contribute to their profits. However, it should help those companies to conduct more efficient clinical trials, by concentrating them on those whose haplotypes suggest they might actually be expected to benefit from a drug. It could also be used to recognise those who would suffer side-effects from a particular drug. Not only will this reduce the cost of testing drugs, it should also increase the number of drugs approved, since they could be licensed only for use by those who would benefit safely. At the moment, only about one molecule out of every ten subjected to clinical trials is licensed. This drop-out rate is a big factor in the cost of getting a molecule to market, which can be as high as $500m.

Little breeders

Having identified your protein, the next question is what you do with it. If you want to turn it straight into a drug, the path is long but well-trodden: trials on cells and tissues, trials on animals, trials on people and, if all that works, eventually an application for regulatory approval.

Several firms, though, are not satisfied with what nature has provided for them. They consider natural proteins mere starting points for drug development, reasoning that a drug, which is treating an abnormal situation (ie, a disease), may thus need to produce an abnormal effect. So they are trying to improve on nature.

All these firms have slightly different versions of the technology, and all, naturally, claim that theirs is the best. But the basic process for each of them is the same: identify several proteins, or several versions of the same protein, that show some of the activity you want; find the genes responsible; divide them into pieces; shuffle the pieces to make a set of new genes, and thus new proteins; set the new proteins to work on the task you are interested in; pick those that work best; then start the whole process over again with those selected genes, and repeat as many times as required to get the result you want.

It is no coincidence that this process precisely replicates natural selection (picking out the best candidates) plus sex (shuffling material between different genes). And it works: at its best, it increases the desired activity more than a thousandfold. It can also be used to make non-medical proteins (of which more later), for which the regulations are less strict, so many of them are already in the marketplace.

Leading molecule-breeding companies working on drugs include

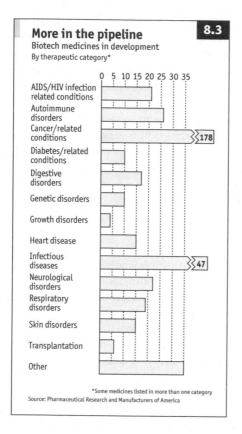

More in the pipeline   `8.3`
Biotech medicines in development
By therapeutic category*

0 5 10 15 20 25 30 35

AIDS/HIV infection
related conditions

Autoimmune
disorders

Cancer/related
conditions ⟩⟩178

Diabetes/related
conditions

Digestive
disorders

Genetic disorders

Growth disorders

Heart disease

Infectious
diseases ⟩⟩47

Neurological
disorders

Respiratory
disorders

Skin disorders

Transplantation

Other

*Some medicines listed in more than one category
Source: Pharmaceutical Research and Manufacturers of America

Applied Molecular Evolution, Genencor and Maxygen. None of their discoveries has yet been approved for use, but there are some interesting prospects around. Applied Molecular Evolution, for example, has an enzyme 250 times more effective than its natural progenitor at breaking down cocaine. Genencor is engaged in designing tumour-killing proteins and proteins that will stimulate the immune system against viruses and cancers, in effect acting as vaccines.

Maxygen is working on improving a group of proteins called interferons. Interferon alpha is currently used to treat multiple sclerosis, but often with indifferent results. Interferon gamma is a remedy for pulmonary fibrosis, an inflammation of the lungs that causes permanent scarring. Maxygen's researchers think they have created more effective versions of both proteins. And, like Genencor, Maxygen is developing proteins it hopes will work as vaccines, mainly for bowel cancer and dengue, an insect-borne fever.

By no means all proteins connected with a disease are appropriate for turning into drugs, but many of those that are not may still have a medical use by acting as targets for drugs. Indeed, the traditional way of doing pharmacology is to find a drug to fit into an active site on a protein molecule, either stimulating the protein, or gumming it up, as appropriate. In the past, such "small molecule" drugs have been identified haphazardly by making zillions of different potential drug molecules, storing them in "libraries" and chucking them at each new protein to see what sticks. However, if you know what a protein looks like, there is an alternative: design an appropriate molecule from scratch.

X-ray specs

Currently the best way of finding out what proteins look like is x-ray crystallography. Pure proteins can frequently be persuaded to crystallise from solution. Fire x-rays through such a crystal and they will interact with the atoms in that crystal. The pattern that emerges will, with enough maths (nowadays done by computer), tell you how the atoms in the crystal are arranged. This works with any crystal, not just one made of protein. Indeed, it was the photographs taken by Rosalind Franklin of the x-ray patterns produced by DNA crystals that gave James Watson and Francis Crick the clue they needed to understand that DNA is a double helix.

x-ray crystallography has already generated several drugs. Viracept, devised by Agouron (now part of Pfizer) and Agenerase, developed by Vertex of Cambridge, Massachusetts, are anti-AIDS drugs that inhibit a protein called HIV-protease. Relenza, devised by Biota Holdings, of Melbourne, Australia, gums up an influenza protein. Until recently, however, x-raying crystals has been a bespoke craft. Now several firms – including Structural GenomiX and Syrrx, both based in San Diego – are trying to industrialise it. They have developed production lines for growing the proteins, and reliable ways of crystallising them. And they are making use of machines called synchrotrons that generate x-rays in bulk by forcing electrons to travel round in circles, which they do not like doing. The resulting screams of protest are beams of x-rays.

Visions of reality

x-ray crystallography has proved effective, but some firms are now looking for more direct ways to determine the properties of a protein. In theory, a protein's shape is implicit in the order of the amino acids in the chain. It is the chemical likes and dislikes of one part of the chain for the other parts that hold the folded molecule together. But to figure all this out takes vast computing power, way beyond the scope of any venture capitalist.

IBM, on the other hand, sees it as a welcome opportunity to put its ever-more-powerful machines to work. Its Blue Gene project is intended to produce a computer that can solve the protein-folding problem. Blue Gene, if it comes to pass, would be a so-called petaflop machine, able to perform a quadrillion calculations a second. The aim was to have something running at a quarter of a petaflop by July 2005. If it happens, it will be a technical tour de force.

Nor are proteins the only bits of biology that can be modelled in a computer. Physiome, a firm based in Princeton, New Jersey, models entire organs. Like nature, Physiome builds these organs from the cell up. Its researchers have developed equations which describe the way cells operate. Each cell then interacts with its neighbours through virtual surface-receptor molecules and ion channels, two classes of protein of interest to drug developers. The result is virtual tissue, and such tissues can be put together to make complete organs.

Physiome's most advanced virtual organ is its heart. This is so lifelike that it responds correctly to electronic hormones and drugs added to its electronic blood supply. The hope is that it and its successors will respond equally accurately to new drug molecules, allowing virtual clinical trials to be run. Such virtual trials would not, of course, be substitutes for real ones. But they would point drug companies towards molecules that were likely to work in real people, making real trials more cost-effective. Cheaper, faster, better drugs are on their way – unless the cash runs out first.

Platform tickets

How to create lots of drugs at once

THE BEST PLATFORM TECHNOLOGIES of all are not those that merely enable drug discovery to proceed, but are new classes of molecules that themselves act as drugs. Therapeutic antibodies are in this category. More than 15 have already been approved by America's Food and Drug Administration (FDA), and well over 100 more are undergoing clinical trials.

Antibodies are the workhorses of the immune system. Like most proteins, they have an "active site" on their surface, which is shaped to fit with part of another molecule. Unlike other proteins, though, antibodies can have active sites of many shapes. In nature, that allows them to lock on to parts of invading pathogens, neutralising the invader. In the laboratory it means that biotechnologists can create antibodies with active sites tailored to perform particular tasks.

One task they are often asked to perform is to attach themselves to a cancer cell. Genentech, the oldest biotechnology company around, has two therapeutic antibodies on the market designed do just that: Herceptin, which attacks breast cancer, and Rituxan, which attacks a form of cancer called non-Hodgkin's lymphoma. The latest wheeze, perfected by IDEC, is to attach a radioactive isotope to an antibody, so that when the isotope decays, the target cell is destroyed by the radiation. This is the most precise form of radiotherapy imaginable. Rheumatoid arthritis is another target. Humira, an antibody developed by Cambridge Antibody Technology, attaches itself to a molecule called tumour necrosis factor which is a vital link in the molecular chain that causes arthritis.

Antibodies were also at the centre of the ImClone scandal. When the FDA rejected the firm's most promising antibody at the end of 2001 because of sloppy clinical trials, Martha Stewart, a lifestyle guru who is a friend of the company's former chief executive, Sam Waksal, was accused of insider dealing. (She was subsequently convicted and spent five months in prison.) Mr Waksal agreed to pay a large fine.

Scandal aside, therapeutic antibodies have proved to be a successful new class of drugs. Isis Pharmaceuticals of Carlsbad, California, has launched another new class which it hopes will be equally successful. Isis controls the critical patents for "antisense" RNA-based drugs. The first of these, designed to take on a bug called cytomegalovirus (which grows in the eyes of AIDS patients, often causing blindness if untreated), is now on sale. Fourteen others, aimed to zap inflammation and cancer, are in the pipeline.

Antisense drugs work by ambushing the messenger RNA molecules that carry the instructions for making proteins from a cell's nucleus to the protein-making machinery outside. These messengers are copies of one of the strands of a DNA molecule in the nucleus. Message-carrying strands are known as "sense" strands, hence their complements are anti-sense.

In principle, RNA molecules can form double-stranded helices, just like DNA does. The reason they do not is that cells do not manufacture the appropriate antisense strands. But such strands can be made by the hand of man. And if sense and antisense should meet, the resulting double-stranded molecule no longer works, so the protein the sense strand encodes is not made any more. If too much of that protein is causing a disease, mugging the messenger this way may stop the illness.

Planting a seed

Despite appearances, agricultural biotechnology has been a success. Whether it will bring truly radical change remains to be seen

MEDICAL BIOTECHNOLOGY may have its troubles, but at least most people favour developing new treatments and methods of diagnosis. Agricultural biotechnology is not so fortunate. Between 1995 and 1998, the area planted with genetically modified crops expanded from nothing to some 30m hectares, mostly in North America. Nobody noticed. Then, after a pointless experiment that involved feeding rats with potatoes modified to produce a poison, parts of Europe developed mass hysteria. In some countries, foodstuffs containing GM ingredients became almost unsaleable.

Matters were made worse by the publication soon afterwards of the results of another experiment, in which pollen from GM maize was fed to caterpillars of the monarch butterfly. The "transgene" that had been introduced was for a natural insecticide called *bt*, and many of the caterpillars died. In the eyes of some (who conveniently forgot to ask what the effect of the insecticidal spray that *bt* replaces would have been), this suggested GM crops damaged the environment. Another worry was that crops containing transgenes might cross-breed with wild plants and produce a generation of superweeds. It did not help when evidence of such escapes was found in Mexico.

In Europe, experimental fields sown with GM crops were duly trashed by environmental activists. Green biotech, evidently, does not appeal to greens. Indeed, paranoia levels became so high that in 2002 some African governments refused food aid that might contain GM grain, in case their own crops were "contaminated" by cross-pollination that would make them unacceptable to European consumers. Rather than risk that, they preferred to let people starve.

It sounds as though agricultural biotechnology is in trouble, but in reality it is not. Though there has been a general downturn in agribusiness recently, sales of GM seeds themselves were worth more than $4 billion in 2002, according to the International Service for the Acquisition of Agri-biotech Applications (ISAAA), which monitors the spread of GM crops. The area planted with genetically modified crops in 2003

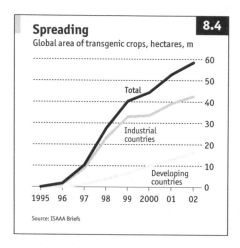

Spreading **8.4**

Global area of transgenic crops, hectares, m

Total

Industrial countries

Developing countries

1995 96 97 98 99 2000 01 02

Source: ISAAA Briefs

amounted to almost 60m hectares – admittedly only 4% of the world's arable land, but a 12% increase on the year before. Where GM strains of a crop species are available, they are starting to dominate plantings of that species. Half the world's soyabean crop is genetically modified. And despite the panic in Africa, three-quarters of those who plant GM crops are farmers in the poor world.

All the same, GM crops have not lived up to the sales patter. In 1996, when such crops started to be introduced in earnest, the market was dominated by just two sorts of modification. One was the addition of *bt* in order to reduce the need for insecticides. The other protected crops against a herbicide called glyphosate, allowing them to be sprayed more effectively. Moreover, only four crops – soya, maize, canola (a high-yielding form of rape) and cotton – accounted for almost all the GM-planted area.

The optimists claimed that this was only the beginning. A few years hence, they said, the world would enjoy better, more nutritious crops, which would be drought-resistant, cold-resistant, salt-resistant and virus-resistant. All this would amount to a new green revolution, courtesy of genetic modification.

In practice, all that has happened is that the protection genetic modification provides against herbicides and insects has been made a bit more effective, and some plants previously protected against one are now protected against both. There have been no genuinely new modifications – which is a pity, because there are a lot of good ideas around, and the poor world in particular could do with them.

Crop circles

It is too easy to blame finicky consumers and Luddite environmentalists for this state of affairs. They have played their part, but the real culprit, as the Nobel-prize-winning Dr Brenner observed, is economics.

The genetics of maize allow the seed market to be controlled by a handful of big firms, including Monsanto, which owns the *bt* and glyphosate-resistance transgene patents. Maize, soya, cotton and canola are supplied to farmers as so-called F1 hybrids, produced by crossing pure parental strains grown exclusively by the seed companies. F1 hybrids do not breed true, so farmers must go back to the seed merchants for new supplies each year. Developing a marketable transgenic strain is almost as costly as developing a new drug, and this kind of control of the market helps to make investing in transgenics worthwhile. Opponents of GM crops who claim that transgenics concentrate power in the hands of seed companies have it backwards. In reality, only those crops already in the hands of such companies have got the treatment.

Another factor is market size. Even if the market is captive, it has to be big enough to justify the investment. Wheat, with a planted area in North America that is about two-thirds that of maize, would be worth genetically modifying. Indeed, Monsanto has developed a glyphosate-protected strain of wheat and hopes to have it approved soon. But transgenic versions of less widely planted crops may never become worthwhile unless the technology gets much cheaper.

Economics also helps to explain why such modifications as have been made are aimed at the farmer rather than the consumer. It has been possible for several years to make seeds containing healthier oils or more vitamins (eg, the famous vitamin-A-enhanced "golden rice"). But such modifications are commercially pointless, at least in the West, where most crops are used in processed foodstuffs rather than sold as raw ingredients. If people want extra vitamins or particular oils, it is easier and cheaper for the food manufacturers to add these in the factory. The retail market for raw ingredients is simply too small to justify spending money on the development and approval of modified versions. And those in the rich world who care about their ingredients might well resist the idea of a new GM strain, however healthy.

Farmers, on the other hand, can see the virtue of paying a bit more for their seed if that allows them to use fewer chemicals. So it is not surprising that the only people interested in using genetic modification to enhance the nutritional qualities of crops are farmers – and they want it for animal feed.

To oblige its customers, Monsanto has formed a joint venture with Cargill, another large agriculture company. This will modify the protein composition of soya and maize grown for animal feed, boosting the levels of essential amino acids (which animals cannot make but have to obtain from their diets). A second deficiency of animal feed, its lack of useful phosphorus, is being tackled by Diversa, a San Diego-based protein-evolution firm. One of its most promising ideas is not a protein-based drug, but an enhanced version of a bacterial enzyme called phytase, which was approved by the American authorities in 2003. Feed contains plenty of phosphorus, but most of it, particularly that in soya, is bound up in a chemical called phytic acid, which mammals cannot digest and which also inhibits the absorption in the gut of trace nutrients such as zinc. Phytase breaks up the acid, liberating the phosphorus and helping micronutrient absorption. That means less need for supplements, and therefore cheaper feed.

A stressful future?

In rich countries where farmers are being paid to take land out of cultivation, improving the resistance of crops to salt, cold and drought is of no great interest, but that has not stopped research completely. For example, Mendel Biotechnology, a small firm based in Hayward, California, has been investigating resistance to such stresses in a plant called *Arabidopsis*, a genetic workhorse that has had its genome completely sequenced. Stress-resistance is known to be controlled by biochemical networks that involve several hundred proteins. Fiddling with these proteins one at a time is unlikely to have much effect, but Chris Somerville, Mendel's boss, thought that the networks might be "tuned" to be more or less active through the use of transcription factors – proteins that regulate the transcription of messenger RNA from genes, and thus control how much protein is produced from a gene.

In partnership with Monsanto and Seminis, the world's largest vegetable-seed firm, Mendel's scientists checked all 1,900 transcription factors produced by *Arabidopsis*. They identified those involved in protecting the plants from salt, cold and drought, and found that altering the expression of those factors could protect the plant more. For example, they produced a strain of *Arabidopsis* that tolerated 17°C of frost. The technique works in crop plants, too. Whether it will ever be commercially viable is a different question. But it might help farmers in the poor world, where drought, in particular, is often a problem. Mendel has offered to donate its drought-protection technology to the Rockefeller Foundation, a large philanthropic organisation, for just that purpose.

Rich farmers, though, might be interested in genes that do the same jobs as existing transgenes, but do them better. To this end researchers at Verdia, a subsidiary of Maxygen, have taken natural fungicidal, insecticidal and herbicide-detoxification genes and improved their efficacy up to a thousandfold. They have even tinkered with rubisco, one of the proteins involved in photosynthesis, and managed to improve its productivity.

Fiddling with photosynthesis would certainly be a radical idea. But biotechnology may also bring radical change of a different sort to farming. Cotton aside, most GM crops are grown for food. Yet white biotechnology could revolutionise the use to which the countryside is put, shifting it away from growing food and towards growing raw materials for industry.

Pharming today

A novel way to make drugs

DOLLY THE SHEEP, the first mammal to be cloned from an adult cell, died in early 2003. Despite the hopes of her inventors at the Roslin Institute in Edinburgh, she did not usher in a new era of animal husbandry. Neither cloning nor genetic modification of commercial farm animals has taken off. Normal breeding, says Harry Griffin, the institute's boss, is more effective, as well as easier.

A handful of firms, though, are pursuing a different take on farming. They are designing animals (and also crop plants) as factories for making therapeutic proteins. Two of these companies lead their respective fields. GTC Biotherapeutics, of Framingham, Massachusetts, is using livestock to make its drugs. Epicyte, based in San Diego, hopes to pull off a similar trick with maize.

GTC's technique is to get its animals to secrete the desired protein into their milk. The gene for the protein in question is inserted into a goat's egg, and to make sure that it is activated only in udder cells, an extra piece of DNA, known as a beta-caseine promoter, is added alongside it. Since beta caseine is made only in udders, so is the introduced protein. GTC now has 15 varieties of engineered goat and is branching out into cows, which have a bigger yield of milk and therefore protein.

Although the firm has yet to bring a product to market, it has high hopes of a substance called antithrombin-3 (AT-3). This is an anti-clotting agent used in coronary-bypass operations and to prevent deep-vein thrombosis in susceptible individuals. A non-caprine version is already approved for use in Europe and Japan, where the combined market is worth $250m a year. GTC hopes to undercut suppliers of those markets if its version of AT-3 passes muster, and also to introduce the drug in America.

Epicyte's researchers have persuaded maize plants to make therapeutic antibodies and express them in large quantities in the endosperms of their seeds, using a promoter-based trick similar to GTC's. This allows the antibodies to be extracted easily, cheaply and in a pure state. The researchers picked maize because the food-processing industry already has a huge amount of experience with this crop. Other products in the works include antibodies against herpes and respiratory syncytial virus, which causes dangerous lung infections in children. In addition, the firm is developing an antibody to one of the proteins involved in Alzheimer's disease. But, like GTC, it has nothing on the market yet.

Besides turning out drugs cheaply (at $1–2 a gram, compared with around $150 a gram from a bioreactor), both of these technologies are easy and inexpensive to scale up. A traditional protein-drug factory costs $200m–400m and takes between three and five years to build. A new strain of goats costs $100m to develop and takes 18 months. And if more capacity is needed, growers can expand quickly by simply breeding more animals or planting more fields.

Reinventing yesterday

Biotech's biggest use may be to rebuild basic industries

ONCE UPON A TIME, much of the man-made world consisted of things that had been grown. Clothes, carpets, bed-sheets and blankets were woven from wool, flax, cotton or (if you were lucky) silk. Shoes were made of leather. Furniture and fittings were made of wood, which also served as fuel for heating and cooking. Then humanity discovered coal, oil and chemistry. Today only the poorest and the richest people burn wood, and many of its other uses have been taken over by plastics. Natural fibres, too, have ceded much of the market to artificial ones. But biology may be about to revenge itself on the synthetic, petroleum-based industrial world by providing new materials and fuels. And in this guise, it may even become acceptable to the environmental movement.

In truth, biotechnology has been quietly working away at industrial applications for some years. It started with enzymes. A business in purifying and selling bacterial enzymes for use in food manufacturing, washing powders and so on has existed for decades, but in 1988 a Danish firm called Novozymes produced the first transgenic enzyme, a fat-digester for detergents. Partly thanks to this lead, Novozymes is now the world's largest enzyme manufacturer, hotly pursued by several other firms.

Enzymes are proteins, which have a reputation for being fussy molecules. Expose them to the wrong temperature, acidity, salinity or pressure and they stop working, sometimes permanently. And the temperature, acidity, salinity and pressure of industrial chemistry is often very different from that found in familiar living organisms. However, it has become clear that lots of bacteria thrive in conditions that used to be regarded as hostile to life. Quite a cottage industry, known as bioprospecting, has developed to collect these bacteria from hot springs, soda lakes, arctic rocks, industrial-effluent outlets and so on. Enzyme companies then analyse the bugs for proteins that look like useful starting points for the sort of directed evolution used by firms such as Applied Molecular Evolution, Genencor and Maxygen in their search for drugs.

Enzyme-catalysed processes have always been a more efficient way of making molecules than traditional chemistry. They often involve fewer synthetic steps, and the yield of each of those steps is almost always close to 100%, whereas the cumulative losses from step to step of doing things in a complicated traditional synthesis mean that the yield may easily end up below 10%. But until recently, the range of reactions for which enzymes could be used was limited, and their fussiness confined them to high-value products such as drugs and vitamins. Now, thanks to directed evolution, there is serious talk of using enzymes to make cheap, bulk chemicals. And not only talk: action, too.

Material progress

The most promising applications for the new model enzymes over the next decade are plastics and fuels. The two most advanced plastics projects are those of DuPont, one of the world's biggest chemical companies, and Cargill Dow, a joint venture between the agricultural and chemical firms of those names. DuPont's process, developed in collaboration with Genencor, took biochemical pathways from three different micro-organisms and assembled them into a single bacterium. The raw material for the process is glucose syrup made from maize starch. This is converted into a molecule called 1,3 propandiol, which is used to make a polyester called Sorona. But Sorona is only half biological. It is a copolymer – that is, it is made out of two sorts of monomer – and the other one, a molecule called terephthalate, still has to be made from oil, so there is some way to go.

Cargill-Dow is closer. Its product, Ingeo, is made out of lactic acid, which in turn is made from glucose. Traditional techniques are used only for the polymerisation of the individual lactic-acid monomers into polylactic acid (the chemical name for Ingeo). The stuff is being made in commercial quantities at a plant in Nebraska, and is about to go on the market. At the moment it is rather more expensive than its petrochemical competitors, but Cargill-Dow hopes to brand it as a premium product in the market for environmentally friendly goods.

Biopolymers are environmentally friendly twice over. Since their manufacture uses little in the way of fossil hydrocarbons, they do not add to global warming. And because they are biodegradable, they cause no pollution when discarded. The firms' bigwigs seem hopeful that this will prove a big enough attraction to allow them to reap economies of scale that will then make their products truly cost-competitive.

DuPont and Dow are giants, but biopolymers can be for minnows too. Metabolix, a small firm based in Cambridge, Massachusetts, takes the process for making them to its logical conclusion – by getting living organisms to do the polymerisation as well as making the monomers.

Animals and plants store surplus energy in the form of carbohydrates, oils and fats. Some bacteria, though, use a different molecule, called a polyhydroxyalkanoate, or PHA. About a decade ago, when they were working at the nearby Whitehead Institute, James Barber and Oliver Peoples, the founders of Metabolix, realised that this material might be put to use as a plastic. They have spent the past ten years proving the point.

Having prospected the bacterial world for appropriate enzymes, and assembled enzymatic pathways in the same way that Genencor did for DuPont, they came up with something new: bugs that actually make plastics and store them inside themselves, in large quantities (about 80% of the weight of a grown bacterium is plastic) and in great variety. PHAS are not a single chemical, but a vast molecular family. Different enzyme pathways can turn out different monomers, producing plastics with different properties. Indeed, it is possible to have two different enzyme pathways within the same bacterium. The result is a copolymer that expands the range of properties still further.

Metabolix has shown that its PHAS, too, can be produced at a price which is competitive with at least the more expensive existing polymers, such as polyesters. That in itself may not be enough to convince manufacturers to switch from tried and trusted materials to Metabolix's novel ones, but the firm hopes that in the large market for single-use items the added feature of biodegradability will be a clincher. If manufacturers do not make the change unprompted, then a nudge from the regulators might be expected. Currently, plastic is a persistent form of rubbish, whereas an object made of PHA will disappear in a few weeks if dumped in a landfill, or even in the sea.

Get the price right, then, and the opportunities are enormous. According to a report published in 2001 by McKinsey, a consultancy, by 2010 biotechnology will be a competitive way of producing about one-fifth of the world's chemical output by value. That means white biotech will be competing in a market worth $280 billion, of which McKinsey thinks the technology might capture about $160 billion. As biotech processes become cheaper, those numbers will increase.

All the companies working in this field have projects designed to bring down the costs. Metabolix, for example, hopes to switch from growing plastics in bacteria (which have to be fed) to growing them in plants (which will make them out of water, carbon dioxide and sunlight). The firm's researchers have already shown that this is possible in the laboratory. They are now scaling up the process.

The enzyme firms, meanwhile, are working on an idea that would allow whole plants to be used as chemical feedstock. Glucose syrup is a refined product, made from maize grains, which form only a small part of the plant. Maize grains cost about $80 a tonne. That is cheaper than petroleum, weight for weight, but the researchers think they can improve on this. Instead of the grains, which are the most valuable part of the plant, they are trying to find ways of using the waste, which fetches only about $30 a tonne for silage. Unfortunately, it consists mainly of cellulose, a natural polymer of glucose but a recalcitrant one.

Help, though, is at hand. The reason dead plants do not stay around indefinitely is that they are eaten by bacteria. These bacteria contain cellulose-digesting enzymes known as cellulases. Genencor and several of its rivals are using this as the starting point for building a better cellulase. Verdia, Maxygen's plant-biotechnology subsidiary, is hoping to go one better. Its researchers are working on developing a cellulase that the plant would make in its own cell walls. To prevent the enzyme digesting the living plant, it would be tweaked to work most effectively in conditions found not inside plants but in bioreactors.

If these ideas come off, then an era of limitless supplies of glucose could follow. That would allow the production not only of as much plastic as anyone could want, but also of another product that can easily be made from glucose: ethanol. This is not only the active ingredient of booze, but also an efficient fuel. Henry Ford's first car was powered by it. Today, some of the motor fuel sold in Brazil is pure ethanol, which modern engines can be tuned to run on happily, and the rest is 20% ethanol. Even in the United States nearly one-tenth of all motor fuel sold is a blend of 90% petrol and 10% ethanol. And since the carbon in ethanol made from plants came out of the atmosphere, putting it back there cannot possibly cause any global warming.

At this point some people in the industry turn starry-eyed and start talking about a future "carbohydrate economy" that might replace the existing "hydrocarbon economy". The countryside would be rejuvenated as a source of raw materials. Land now taken out of cultivation would be put back to use. Small-scale chemical plants to process the stuff would pop up everywhere. And the oil-producing countries would find themselves out of a job.

Surprisingly, these visionaries are often hard-headed businessmen. Even more surprisingly, the numbers they are bandying do not sound all that exotic. The American market for bioethanol is already 8 billion litres a year (see Chart 8.5). The enthusiasts at Genencor reckon it could be as high as 75 billion litres a year by 2020. That would be enough to replace two-thirds of America's current petrol production. In January, a Canadian firm called Iogen opened a small cellulase-powered pilot plant that converts straw into ethanol.

The germ of an idea

Finding enzymes such as cellulases involves, as mentioned earlier, bioprospecting. But there is bioprospecting, and then there is Craig Venter. Dr Venter was the man behind Celera, the company that took on the Human Genome Consortium. The firm got its scientific edge from a technique called whole-genome shotgun sequencing, which he had developed to work out the genetic sequences of bacteria in one go. Using the money Celera had raised, Dr Venter applied the technique to the much more complicated task of working out the human genome in one go. Now, he proposes to apply it to entire ecosystems, working out the genomes of all the critters in them by a similar, one-step approach. Admittedly the critters are bacteria, and the ecosystems are water samples from the Sargasso Sea. But such samples will have thousands of species in them, most of which cannot be cultured in the laboratory and are therefore inaccessible to standard sequencing methods.

Whole-genome shotgunning works by shredding the DNA of an organism into tiny pieces, sequencing the pieces, then sticking the results together again in the right order, using a powerful computer and clever software. Whole-ecosystem shotgunning aims to do the same with all the DNA in a sample, regardless of how many species it comes from. If the software is good enough, it will be able to sort the pieces into the individual genomes.

Dr Venter is full of ideas about what might be done with his discoveries, even before he has made them. But he is particularly excited by the possibilities for energy generation, and set up an organisation, the Institute for Biological Energy Alternatives, to investigate them further. In his view, replacing petrol with ethanol is old-think. New-think would power the world not with internal combustion engines but with fuel cells. And fuel cells use hydrogen.

One way to make hydrogen biotechnologically might be with a bug called *Carboxydothermus*, which was discovered in a hydrothermal vent (an undersea volcanic spring) off the coast of Russia. This species lives by reacting carbon monoxide with water. One of the waste products is hydrogen. A more promising route might be to intercept the hydrogen ions produced in the first step of photosynthesis. Another of Dr Venter's pet projects, creating a bacterium with a completely synthetic genome, could come into its own here. By leaving out the genes for the sugar-forming pathways that normally use these hydrogen ions, such a creature could be made to devote all its energies to producing

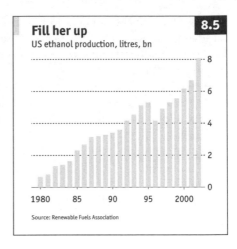

Fill her up 8.5

US ethanol production, litres, bn

Source: Renewable Fuels Association

hydrogen. Nor could it escape into the outside world (always a worry with bio-engineered bugs), because it would lack the biochemical apparatus to survive there. Thus trapped, it could, he muses, be used in solar-powered fuel cells for such applications as portable computers.

That points to the power of industrial biotechnology to create completely new products. The idea of a partly living fuel cell may be merely dipping a toe in the ocean of possibilities. Another dipped toe is that of Nexia Biotechnologies, based in Quebec, which is using technology similar to that of GTC Biotherapeutics to turn out spider silk in goats' milk. Spiders, observes Jeffrey Turner, the firm's boss, have been perfecting silk for the past 400m years. Such silk comes in many varieties, which do different jobs for spiders and can thus do different jobs for people. For Nexia's products, these jobs range from stopping bullets when the silk is worn as body armour to stitching up eyeballs after surgery.

Some firms, such as Genencor, are starting to explore wilder shores. As proponents of nanotechnology (the incipient field of building devices a few billionths of a metre across) are wont to observe, biology is natural nanotechnology. Why, then, go to all the trouble of creating an artificial nanotechnology from scratch? Genencor is collaborating with Dow Corning, a big materials company, in this area. Among other things, the firms are looking at rhodopsins, the protein molecules that act as light-detecting pigments in a range of organisms from bacteria to people. Genencor has bred 21 rhodopsin-type molecules, each of which responds to a particular wavelength of light. These molecules might have applications as switches in photonics, the as yet largely hypotheti-cal idea that data could be processed by light instead of by electronics. These are small-scale investments, and may come to nothing, but they are worth a flutter.

However, there are shores yet wilder than these awaiting, where big battles are almost guaranteed. For among the prospects offered by biotechnology is one hitherto reserved for science fiction: tailor-made humans.

The bugs of war

Almost certainly on their way

SOONER OR LATER, almost every new technology is pressed into service in warfare, and that is likely to go for biotechnology too. In one sense, that is trivial. As biotechnology, like chemistry before it, yields new materials, fuels and other industrial products, these will be used to make military matériel. But biotechnology, again like chemistry, may also yield new weapons.

Most civilised countries have signed conventions that prohibit them from deploying – indeed, from manufacturing – biological weapons. The same is true of chemicals, and civilised countries do not use chemical weapons either.

But uncivilised countries do, and so do terrorists. If they thought biological weapons served their cause, they would probably use them, too – especially if biotechnology helped to make such weapons far more effective than they are now.

This is an area that people are understandably reluctant to talk about, but the most terrifying act possible short of letting off a nuclear weapon would probably be to spread a lethal plague. At the moment, it is generally agreed that smallpox has the best mixture of transmissibility and deadliness to do the job effectively. But a better bug might be engineered.

Putting a viral genome together from scratch would not be that hard or expensive (perhaps $1m), and it could probably be done without anyone noticing. Many companies sell short stretches of DNA and RNA, called oligomers, for research purposes. In principle, any viral genome can be stitched together from such oligomers. Indeed, in 2002 Eckard Wimmer and his colleagues at the University of New York's Stony Brook campus announced they had created a polio virus from scratch this way.

If orders for the oligomers were spread between, say, a dozen different firms, no one would spot what was going on. A group of scientists could then "custom build" a viral genome, picking genes from different natural viruses for different properties, such as virulence and transmissibility. The genetic sequences of many such viruses are available in public databases, and more are being added all the time.

In America, the most likely target of such terrorist ire, people are starting to think about the problem. Better and faster gathering of information on symptoms that people are reporting to their doctors might allow an epidemic to be detected early, but if it was caused by a previously unknown agent that would be no help in treating it. And new vaccines currently take years to develop. Whether biotechnology can reduce that time is moot.

Man and superman

Biotechnology could transform humanity – provided humanity wishes to be transformed

WARNING AGAINST INTELLECTUAL ARROGANCE, Alexander Pope wrote: "Know then thyself, presume not God to scan; the proper study of mankind is man." But his words have turned out to be misguided. Though studying man may not exactly have led scientists to scan God, it has certainly led to accusations that they are usurping His role.

More drugs; cheaper food; environmentally friendly industry. Who could object? But people do. The image that haunts biotechnology, and perhaps the most influential piece of science fiction ever written, is Mary Shelley's *Frankenstein*. When the book was first published in 1818, most people did indeed believe that life was created by God. Shelley's student doctor apes that act of divine creation and comes a cropper. He has come to epitomise the mad-scientist figure: either downright wicked, or at the least heedless of humanity's good.

The book's subtitle, though, is telling: *The Modern Prometheus*. Prometheus, in the Greek myth, stole fire from heaven and gave it to mankind with the intention of doing good. The reason Prometheus was punished by his particular set of gods was that he gave mankind power, and with that power, choice.

Biotechnology is not about to create a human from off-the-shelf chemicals, nor even from spare parts. But it may soon have the power to manipulate human life in ways which could bring benefits, but which many will find uncomfortable or abhorrent. A choice will have to be made.

Clones to the left of me ...

No one has yet cloned a person, or genetically modified one, at least a whole one. But people are working on technologies that could help to do these things.

An existing individual might be cloned in several ways. The first would be to persuade a cell (say a skin cell) from the individual to be cloned that it was, in fact, a fertilised egg. That would mean reactivating a whole lot of genes that skin cells don't need but eggs do. As yet, no one knows how to go about that.

The second way is the Dolly-the-sheep method, which is to extract the nucleus of an adult cell and stick it in an egg from which the nucleus has been removed. That seems to trigger the desired reprogramming. Or instead of putting the nucleus into an egg cell, it might be put into a so-called stem cell from an early embryo. Embryonic stem cells can turn into any other sort of cell, so might possibly be persuaded to turn into entire people.

Regardless of that possibility, embryonic stem cells have medical promise, and several firms are currently studying them. Geron, the most advanced of these firms, has worked out how to persuade embryonic stem cells to turn into seven different types of normal cell line that it hopes can be used to repair damaged tissue. Blood cells could be grown in bulk for transfusions. Heart-muscle cells might help those with coronary disease. "Islet" insulin-secreting cells could treat diabetes. Bone-forming cells would combat osteoarthritis. A particular type of nerve cell may help sufferers from Parkinson's disease. Cells called oligodendrocytes may even help to repair the insulating sheaths of nerve cells in people with spinal injuries. Geron is also working on liver cells. In the first instance, these would be used not to treat people, but to test potential drugs for toxicity, because most drugs are broken down in the liver.

Such transplanted tissues might be seen as foreign by the immune system, but Geron is keeping its corporate fingers crossed that this can be dealt with. Embryos have ways of gulling immune systems to stop themselves being rejected by the womb. In case that does not work, though, the discussion has turned to the idea of transplanting adult nuclei into embryonic stem cells as a way of getting round the rejection problem. This idea, known in the trade as therapeutic cloning, has caused alarm bells to go off. The technique would create organs, not people, and no one yet knows whether it would work. But some countries are getting nervous about stem-cell research. This nervousness has not been calmed by the activities of Advanced Cell Technology, a firm based in Worcester, Massachusetts, which announced in November 2001 that it had managed the trick of transplanting adult nuclei into stem cells and persuading the result to divide a few times. In effect, ACT created the beginning of an embryo.

In 2002 President George Bush issued a decree restricting federal funding in America to existing embryonic stem-cell lines. Attempts have even been made in Congress to ban it altogether. Reversing the usual traffic flow, some American scientists have upped sticks and gone to Britain, where the regulations on such research are liberal and settled.

Some countries, indeed, have more than just settled regulations. Singapore, for example, is actively recruiting people who want to work on the human aspects of biotechnology. China, too, is said to be interested. Cynics might regard this as opportunism. But not everyone's moral code is shaped by Judeo-Christian ethics – and besides, moral codes can change.

At the moment, cloning mammals is a hazardous business. It usually requires several hundred attempts to get a clone, and the resulting animal is frequently unhealthy, probably because the original transplanted nucleus has been inadequately reprogrammed. Nor does there seem to be much of a market, so no one is trying very hard.

Genetic modification is a different matter. GTC's drug-producing and Nexia's silk-producing goats are valuable, and people are putting in serious work on the technology. If someone wanted to add the odd gene or two to a human egg, they could probably do so. Indeed, something quite similar is already being done, although under another name: gene therapy intended to deal with illnesses such as cystic fibrosis is in fact a type of genetic modification, although admittedly one that is not passed from parent to offspring. But extending gene therapy to germ cells, to stop the disease being passed on, is under discussion.

... jokers to the right?

A scene in *Blade Runner*, a film that asks Shelleyesque questions about the nature of humanity, is set in the headquarters of a prosperous-looking biotechnology company. The firm makes "replicants", robots that look like humans, and the firm's boss describes how they are grown from a single cell. The replicants, it is plain, are genetically modified people without any legal rights. In this dystopia, it is the unaltered humans who rule. By contrast, GATTACA, another movie set in a genetically modified future, has the modified in charge. They are beautiful, gifted and intelligent. It is those who remain untouched by modification who suffer.

All this is in the realm of fiction, but the contrasting views of the potential effects of genetic modification point to an important truth about any technology. What really matters is not what is possible, but what people make of those possibilities. In the fantasy worlds of science fiction, people are frequently dominated by the technology they have created, and made miserable as a result. Yet so far, the real technological future ushered in by the industrial revolution has defied the fantasists. Dystopia has failed to materialise.

Perhaps, one day, some tyrant will try to breed a race of replicant slaves, but it seems unlikely. It seems much safer to predict that the rich will attempt to buy themselves and their children genetic privileges if and when these become available. But there is nothing new in the rich trying to buy privileges. The antidote is not a Draconian ban on basic research, but reliance on the normal checks and balances, both legal and social, of a liberal society. These have worked in the past, and seem likely to work in the future.

Tyranny, by definition, is incompatible with liberalism. More subtly, the one near-universal feature of technologies in liberal societies is that in time popular ones get cheaper as market competition does its work. Personal genetic modification may start out aristocratic, but if it does turn out to be a good thing, it will become demotic. Conceivably, it may indeed prove to be the field's killer application. And perhaps it is a useful antidote to hysteria to point out that trite, fun applications – say, temporarily changing your skin pigmentation – are conceivable, too.

Critics may say that decisions on cloning and germ-line modification are different, because they affect an unborn individual who has no say in the matter. But equivalent decisions about the unborn are routinely made already, albeit with the watchful eye of the law firmly on the decision-maker.

Even if people do not choose to alter themselves, though, biotechnology is likely to become ubiquitous. Its potential is too great to neglect. Its current woes will come to be seen as mere teething troubles. The first route to ubiquity is likely to be via the chemical industry. As people become more confident about manipulating enzymes and micro-organisms, ever larger swathes of industrial chemistry will fall into biotech's grip. Like existing chemistry, though, the results will be taken for granted almost instantly.

Health care will also be revolutionised by biotech: not merely through new drugs, but through the ability to deploy them precisely and to anticipate the need for their use from studies of an individual's haplotype. Medicine will become less of an art and more of a science. It may even become a consumer good, if drugs intended to let people operate beyond their natural capacities are developed. That, though, is another area fraught with moral difficulties.

What remains unclear is the extent to which bioengineered organisms will become products in their own right. The raspberry blown at GM crops, which are the only transgenic species on the market at the moment, does not encourage the idea that modified organisms will be welcomed with open arms. But captive, genetically modified micro-organisms, such as those that would run Dr Venter's putative solar-powered fuel cells, probably do have a big future.

Large organisms, too, may be exploited in ways as yet hard to imagine: furniture that is grown, rather than made; clothing that eats the dead skin its wearer sheds; miniature pet dragons (fire-breathing optional) as household pets. Whatever happens, however, it will be because somebody wants it to. Bacon was right. Knowledge is power – and generally a power for good. The century of Watson and Crick is just beginning.

The material on pages 236–71 first appeared in a survey in *The Economist* in March 2003.

9

ENERGY

These fuelish things

The fuel cell is enchanting politicians on both sides of the Atlantic. It is too soon, though, for them to dream of freedom from fossil fuels

WHERE IN THE WORLD can you find hydrogen? At first blush, that might seem a ridiculous question: hydrogen, after all, is the commonest element in the universe. The problem is that it is rarely found in its free state on earth. If you want to get your hands on some hydrogen, you generally have to strip it away from carbon, as found in hydrocarbon fuels, or from oxygen, as found in water. Either way, energy is required to produce it. And that, in a nutshell, is the big drawback lurking behind all the recent hoopla surrounding the charms of hydrogen energy.

The hoopla began at the end of 2002, when the European Commission unveiled a grand, €2.1 billion ($2 billion) "hydrogen vision". Romano Prodi, then the commission's president, even declared that he wanted to be remembered for only two things: the European Union's eastward expansion, and hydrogen energy. In 2003 George Bush, America's president, produced his own $1.2 billion hydrogen plan (he even examined a hydrogen-powered car, and made sure he was photographed doing so). In speeches directed at the car industry in Detroit and at the oil industry in Houston, Mr Bush and his team made the claim that the rise of the fuel cell would consign the internal-combustion engine to the dustbin of history. And if that were not enough, Democratic rivals in Congress – trying to keep up – unveiled their own hydrogen initiative.

Fuel cells are devices that work rather like batteries, converting chemical energy into electricity and heat. All fuel cells combine hydrogen with oxygen to produce power. These nifty power plants can be used to run anything from a mobile phone to an office complex. Their greatest attraction is that they can do all this without generating emissions any more harmful than water vapour.

The catch, of course, is that it is first necessary to find a source of hydrogen. If renewable energy is used to split water into hydrogen and oxygen by electrolysis, then the energy produced by a fuel cell is genuinely emission-free. But if energy from a hydrocarbon such as petrol or coal is used, there will still be some unwanted emissions. That applies even if the route taken is steam reformation, in which the hydrocarbon is reacted with water vapour to liberate the hydrogen in both, rather than being used to make electricity for the electrolysis of water.

The emissions from steam reformation, though, are less than those created when the same amount of hydrocarbon is burned in today's combustion engines. This is because fuel cells produce electricity

Ballooning savings	9.1

The cost of producing hydrogen

	$ per gigajoule
Electricity from nuclear power	10–12
Electricity from coal/gas minus CO_2	15–18
Hydrogen from coal/gas/oil	1–5
Hydrogen from natural gas minus CO_2	8–10
Hydrogen from coal minus CO_2	10–13
Hydrogen from biomass	12–18
Hydrogen from nuclear power	15–20
Hydrogen from onshore wind	15–25
Hydrogen from offshore wind	20–30
Hydrogen from solar cells	25–50

Source: International Energy Agency

efficiently, without combustion. And, if techniques for capturing and "sequestering" the carbon dioxide produced by hydrocarbons are perfected, it would make hydrogen from fossil fuels a great deal cleaner still.

How the ghost of you clings

Europe and America do not see eye to eye on the question of how best to generate hydrogen. Europe is putting more emphasis on renewables; America, by contrast, is keen on the possibility of deriving hydrogen from fossil fuels.

At the moment, using renewables is an expensive way of generating hydrogen (see Chart 9.1). So why is Europe heading in this direction? Alessandro Ovi, one of Mr Prodi's advisers, explained that Europe's push for hydrogen is motivated largely by a desire to meet its commitments to cut greenhouse gases under the Kyoto treaty on global warming. Accordingly, the EU has adopted demanding targets for increasing the share of renewable energy to 22% of the region's electricity supply by 2010, up from about half that in 2003.

Such a target for renewable energy sounds pretty green, but there is a snag: wind and solar energy are intermittent, and unlike other commodities – be they bananas or natural gas – there is no good way to store electricity for later use. No way, that is, unless you use renewable energy to produce hydrogen, and store this instead. It can then be used when the power grid is facing peak demand and the price of energy thus increases. Dr Ovi thinks hydrogen could transform the economics of renewables and play an essential role in the EU's clean-energy strategy.

Mr Bush's plan pushes instead for hydrogen via fossil fuels, because greenery is not the only attraction of fuel cells. Mr Bush insists that hydrogen is a good way to bolster his country's "energy independence" from Middle Eastern oil. Hydrogen can be made from America's plentiful supplies of coal, as well as from locally produced biomass and renewable energy, says John Marburger, Mr Bush's top science adviser. Thus, America's reliance on oil from fickle foreign regimes will decline. That vision of energy independence through fossil hydrogen is also gaining popularity among the leadership in coal-rich but oil-starved China.

Does that mean the American approach is ungreen? Not necessarily. Even if fossil fuels were used to produce hydrogen without sequestration, fuel-cell-powered cars would still produce zero local emissions on roads. (Wags call this "drive here, pollute elsewhere".) Further, hydrogen is likely to be produced by some green sources anyway: in the Pacific north-west, hydroelectric power is dirt cheap at night, and on the windswept Great Plains renewables or biomass may prove more economic than fossil fuels.

If America pursues its hydrogen vision by using fossil fuels with techniques such as sequestration, a technology Mr Bush has repeatedly applauded, its hydrogen embrace will indeed be greener than green. What is more, if Big Oil also gets behind hydrogen – as it is now starting to do thanks to the push from the Texan oilman in the White House – the thorny question of where you can find hydrogen could one day become very simple to answer. Right at your corner petrol station.

The material on pages 274–6 first appeared in *The Economist* in February 2003.

Batteries not included?

Will tiny versions of the fuel cells now being developed for cars soon power laptop computers too?

As video telephony, broadband internet links and other high-powered features are added to laptop computers, personal digital assistants (PDAs) and mobile phones over the next few years, the energy demands of these devices will soar. The Samsung Advanced Institute of Technology, the research arm of the *chaebol* of that name, estimates that such upgraded portable devices will require power sources with at least 500 watt-hours per kilogram of energy stored in them. Lithium-ion batteries, today's best, can manage half that, but even the most optimistic estimates suggest that only a 30% improvement could be squeezed out of such batteries.

But there may be an alternative. Miniature fuel cells, which generate electricity by reacting hydrogen with oxygen, can do much better than batteries – at least in a laboratory. The question is whether they can ever do so in the real world. This was the subject of a conference organised in May 2003 in New Orleans by the Knowledge Foundation.

It's not a gas

The key to making fuel cells small is to replace the hydrogen – or, rather, to deliver it in a non-gaseous form, since it is hardly practical to fit portable electronic devices with pressurised cylinders. In the long run, there may be ways round this, for instance by absorbing the gas in metal hydrides or carbon nanotubes. But in the short term the solution seems to be to deliver the hydrogen as part of a hydrogen-rich compound, such as methanol. This is a liquid, which means it is easy to handle. Sachets of methanol fuel, purchased at newspaper kiosks, rather like refills for cigarette lighters, could be inserted with little fuss into electronic devices.

There are two ways to get the hydrogen out of methanol in a way that a fuel cell can use. One, being pursued by several companies, notably Motorola, is called reformation. This attempts to replicate in miniature the complicated networks of piping, heaters, vaporisers, heat exchangers and insulation that the petrochemical industry uses to extract hydrogen in bulk from methane, a chemical one oxygen atom different from methanol. That is hard – doubly so, since reformation works best at 200°C.

There are some variations on the theme, but most of those building miniature methanol reformers use an approach not much different from the one used to make circuit boards for computers. They laser-drill holes into tiny ceramic wafers to guide the flow of fluids. Then they stack these one on top of another, like layers of a sandwich, sinter them together at a temperature of 800°C, and laminate them. Presto, a mini-chemical plant.

The alternative to reformation is to feed the methanol directly into the cell, and rely on a catalyst to break it up at the electrodes, where the hydrogen is separated into its constituent electrons (which form the current that the cell produces) and protons. The trouble with this approach is that pure methanol tends to get everywhere, and thus wrecks the cell. Diluting it with water reduces this problem, but also reduces the power output.

At least one firm, however, thinks it can get round this. MTI Micro-Fuel Cells, based in Albany, New York, boasts some top researchers poached from the Los Alamos National Laboratory in New Mexico. One of them, Shimson Gottesfeld, told the conference that the firm has developed a cell that can use undiluted methanol. This, he claims, allows it to achieve more than three times the energy density of lithium-ion batteries.

Dr Gottesfeld was reluctant to go into details. But the secret seems to lie in some clever internal geometry, which eliminates the need for pumps. That, in turn, reduces the tendency for methanol to go where it is not wanted. However the trick is performed, though, MTI has working prototypes. It also has a contract. Its cells went on sale in 2004 as part of a hybrid power-pack (that is, one which also involves batteries) built by a large equipment firm called Intermec for use in handheld computers.

Better ways of handling methanol are not the only possibility, though. Another is to find further alternatives to elemental hydrogen. That is the route chosen by Medis Technologies, an Israeli-American firm. Its fuel is a mixture of glycerol and sodium borohydride. These react in the presence of a platinum and cobalt catalyst, generating protons and electrons in the same way as methanol – or, indeed, pure hydrogen.

Although many at the conference were sceptical, suggesting for example that the Medis cell works only when it is standing up, the firm remains bullish. Gennadi Finkelshtain, Medis's principal scientist, acknowledges his device's sensitivity to its orientation, but insists that he has a solution in the works. The fact that he has persuaded General Dynamics, a big defence contractor, to form a partnership with him to supply the American armed forces with the new device suggests that the problems cannot be too great. Medis has already demonstrated a prototype recharger for a "ruggedised" military PDA.

None of this adds up to a revolution in portable power, of course, but it is a tantalising start. As so often with new technologies, military applications are important drivers. The American army is keen to have more energetic and longer-lived power sources for such things as climate-controlled bodysuits, advanced mobile-communications equipment and more sophisticated sensors. But consumers will soon be able to purchase lightweight fuel cells, too. Even though they are unlikely ever to be compact enough for use in mobile phones, they could act as portable chargers for such phones. And in devices that are only a bit larger, they could, indeed, end up replacing batteries.

The material on pages 277–9 first appeared in *The Economist* in May 2003.

Hooked on lithium

Without the lithium-ion battery, introduced over a decade ago, portable gadgets – from mobile phones and video cameras to laptops and palmtops – would have remained brick-like objects best left on the desk or at home. But the innovation would have floundered had electro-chemists in America not teamed up with a Japanese firm

THE MOBIRA SENATOR, launched in 1982 by Nokia, was the grand-daddy of today's mobile phones. It consisted of a small handset connected to a brick-like battery pack, with a hefty handle on top – a vital feature, since the whole thing weighed 9.8kg. Today, a typical mobile phone is a hundredth of this (ie, 100 grams or less) and can be tucked discreetly into a shirt pocket. This 99% weight reduction has been achieved largely through advances in battery technology. Above all, it is down to one particular breakthrough: the advent of the lithium-ion rechargeable battery.

Lithium-ion batteries are the foot-soldiers of the digital revolution. They power telephones, music players, digital cameras and laptops. They are amazingly small and light, and can store more energy in less space than any other type of rechargeable battery. A modern handset can run for several days on a lithium-ion battery the size of ten business cards stacked on top of one another. A nickel-metal-hydride battery of equivalent capacity would weigh twice as much and be about twice as big; nickel-cadmium and lead-acid batteries are heavier and larger still.

Lithium-ion batteries are also superior in that they do not suffer from "battery memory effect", a loss of capacity that occurs when a battery is recharged before it is fully depleted. As a result, lithium-ion batteries now account for 63% of worldwide sales of portable batteries, compared with 23% for nickel-cadmium and 14% for nickel-metal-hydride.

But getting lithium-battery technology from the laboratory into millions of people's pockets was not easy, and took many years. Engineers had numerous problems to overcome. Having solved them, they are now working on even more advanced lithium batteries that will be more versatile still. Unless something entirely new comes along – microscopic fuel cells, for example, or more efficient photo-voltaic cells – the world's gadgets are likely to be running on lithium for some time to come.

Rechargeable batteries work by exploiting reversible electro-chemical reactions at the battery's positive and negative terminals (electrodes). The battery is charged by applying an electric current, which causes a reaction that puts the battery into a high-energy state. The battery can then be used as a power source. As it discharges, the electro-chemical reaction operates in reverse, releasing energy and returning the battery to its uncharged, low-energy state. Different battery types, such as lead-acid and nickel-cadmium, use different compounds and chemical reactions, but the basic principle is the same for each.

Lithium has particular appeal for use in batteries for two reasons, observes Michel Armand, a researcher at the University of Montreal and co-author of a paper in *Nature* about lithium batteries. First, it is the most electro-positive metal, which means it can be used to make batteries with higher terminal voltages (typically, 4 volts rather than 1.5 volts) than other designs. And, second, it is the lightest metal, having the capacity to store 3,860 amp-hours of charge per kilogram of weight compared with 260 amp-hours per kilogram for lead.

This means that lithium-based batteries have a high energy density, enabling them to pack a lot of energy into a small, light package. Accordingly, lithium batteries have been used since the 1970s to power watches, calculators and medical implants. But such batteries are non-rechargeable. Making a lithium battery that could be recharged was far from easy.

One early attempt, a research project launched in 1972 by Exxon, an American energy giant, used titanium sulphide as the positive electrode and pure lithium metal as the negative electrode. Titanium sulphide was chosen because it is an "intercalation" compound – a substance with a layered, crystalline structure that can absorb other particles between its layers, much as a sponge soaks up water. Discharging the battery causes charged atoms (ions) of lithium to break away from the negative electrode, swim through an intermediate liquid (electrolyte) and take up residence within the titanium-sulphide lattice, releasing energy in the process. Recharging the battery causes the lithium ions to swim back again, and to reattach themselves to the negative electrode.

The drawback with this design, however, is that after repeated charge/discharge cycles the lithium does not form a perfectly smooth metal at the negative electrode. Instead, it assumes an uneven, "dendritic" form that is unstable and reactive, and may lead to explosions. This problem has never been completely solved, says Frank McLarnon, a researcher at the Lawrence Berkeley National Laboratory in California. Replacing the negative electrode with a lithium-aluminium alloy reduced the risk of explosion, but made the battery far less efficient and long-lived.

Rocking on

Instead, a new approach was tried, and was working well in the laboratory by the late 1980s. It involved using a second intercalation compound to play host to the lithium ions at the negative electrode. Discharging and recharging the battery then simply caused the lithium ions to move back and forth between the two intercalation compounds, releasing energy in one direction and absorbing it in the other. This to-and-fro technique became known as "rocking-chair" technology. The use of lithium in its ionic, rather than its metallic, state made the batteries much safer than previous designs – and explains why they are known as "lithium-ion" batteries.

By this time, a team of researchers led by John Goodenough, now at the University of Texas, had discovered a new family of intercalation compounds based on oxides of manganese, cobalt and nickel. The first commercial lithium-ion battery, launched by Sony in 1991, was a rocking-chair design that used lithium-cobalt-oxide for the positive electrode, and graphite (carbon) for the negative one.

Charging the battery causes lithium ions to move out of the cobalt-oxide lattice and slip between the sheets of carbon atoms in the graphite electrode – a state of higher potential energy. Discharging the battery causes them to move back again, releasing energy in the process (see Chart 9.2). This type of battery is now in widespread use, and Dr Goodenough was awarded the $450,000 Japan Prize in 2001 in recognition of his work.

As well as identifying the right combination of materials, Sony's key contribution, says Dr McLarnon, was to recognise that the battery must not be overcharged. If too many lithium ions are extracted from the cobalt-oxide lattice, it will disintegrate. Besides, packing too many lithium ions into the graphite lattice can result in small particles of lithium being formed, with possibly dangerous consequences. Sony implemented a number of safety measures, including the use of a porous polymer that melts together if the battery overheats, thus preventing ion transport and shutting the battery down. It also developed a

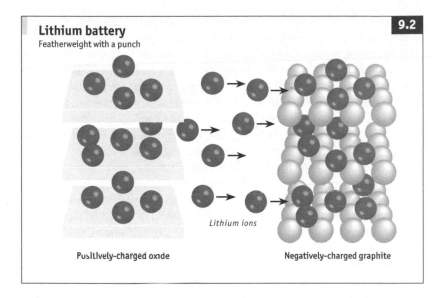

Lithium battery
Featherweight with a punch

9.2

Lithium ions

Positively-charged oxide

Negatively-charged graphite

special "smart" charger with software that prevents overcharging. Such chargers are now commonplace.

Since the commercialisation of lithium-ion batteries began in the 1990s, there have been a few minor improvements, and the cost has come down substantially. One clever tweak, which has made possible the ultra-thin batteries used to power mobile phones, was the replacement of the liquid electrolyte between the battery's two electrodes with a porous separator soaked with an electrolytic gel. This means that the battery can be put together like a sandwich, making thin, flat designs possible.

The next step is to move from a gel to a solid polymer electrolyte, which will be thinner and lighter still. The problem, says Dr McLarnon, is that lithium ions diffuse far more slowly through a solid than through a gel or liquid, making the battery slow to charge and discharge – unless the polymer is made extremely thin.

Still in its infancy

After that? According to Dr Armand, rechargeable lithium-based battery technologies are still in their infancy, and there is much scope for improvement. A number of promising new materials are being developed, including new intercalation compounds and aerogels for the positive electrode, as well as novel alloys for the negative electrode. Some researchers have been working on new ways to prevent dendritic growth in lithium metal-based batteries, using special electrolytes.

If lithium-ion batteries are so light and powerful, why not use them in electric cars? The main problem, says Dr McLarnon, who is working in this area, is cost. A set of lithium-ion batteries capable of powering an electric car now costs around $10,000. Lead-acid batteries may be five times heavier, four times bulkier and far less efficient – the General Motors EV-1, for example, is powered by 533kg of lead-acid batteries that take eight hours to recharge – but they are also a lot cheaper.

But now that all-electric cars have fallen out of favour, since most of their alleged benefits can also be achieved by hybrid (petrol/electric) vehicles, lithium-ion is back in the running. Hybrid vehicles need a much smaller power pack, bringing the cost of a suitable lithium-ion battery down to below $1,000. Even so, says Dr McLarnon, there is still work to be done. The batteries may wear out after a few years of use, and they must be optimised to deliver, and absorb, sudden bursts of power, rather than large amounts of energy at a roughly constant rate, which is what consumer-electronics devices require. Dr McLarnon is optimistic that this can be achieved.

The development of the lithium-ion battery is an object lesson in how pure and applied research, driven by commercial interests, can generate the incremental improvements in a technology that are necessary for transforming it into a useful product. In this case, intercalation compounds were an offshoot of pure research into superconductivity. They were then picked up by Dr Goodenough and other researchers working on battery technology; and the final pieces of the puzzle were supplied by Sony. (Dr Goodenough, who did his original research at Oxford, says battery firms in the West rejected his approaches.)

There was no single "eureka" moment, but a series of gradual improvements – with the baton passed between a number of different groups. "These things zigzag back and forth," says Dr Goodenough. "That's how innovation works." The baton has now been passed to new researchers seeking further improvements and applications, from video-phones to cars. Lithium-ion batteries, it seems, still have a long way to go.

The material on pages 280–4 first appeared in *The Economist* in June 2002.

Building the energy internet

More and bigger blackouts lie ahead, unless today's dumb electricity grid can be transformed into a smart, responsive and self-healing digital network – in short, an "energy internet"

"TREES OR TERRORISTS, the power grid will go down again!" That chilling forecast comes not from some ill-informed gloom-monger or armchair pundit, but from Robert Schainker, a leading expert on the matter. He and his colleagues at the Electric Power Research Institute (EPRI), the official research arm of America's power utilities, are convinced that the big grid failures of 2003 – such as the one that plunged some 50m Americans and Canadians into darkness in August, and another a few weeks later that blacked out all of Italy – were not flukes. Rather, they and other experts argue, they are harbingers of worse to come.

The chief reason for concern is not what the industry calls "poor vegetation management", even though both of 2003's big power cuts were precipitated by mischievous trees. It will never be possible to prevent natural forces from affecting power lines. The real test of any network's resilience is how quickly and intelligently it can handle such disruptions. Think, for example, of the internet's ability to reroute packets of data swiftly and efficiently when a network link fails.

The analogy is not lost on the energy industry. Of course, the power grid will never quite become the internet – it is impossible to packet-switch power. Even so, transforming today's centralised, dumb power grid into something closer to a smart, distributed network will be necessary to provide a reliable power supply – and to make possible innovative new energy services. Energy visionaries imagine a "self-healing" grid with real-time sensors and "plug and play" software that can allow scattered generators or energy-storage devices to attach to it. In other words, an energy internet.

Flying blind

It sounds great. But in reality, most power grids are based on 1950s technology, with sketchy communications and antiquated control systems. The investigation into 2003's North American blackout revealed that during the precious minutes following the first outages in Ohio, when action might have been taken to prevent the blackout spreading, the

The shape of grids to come?

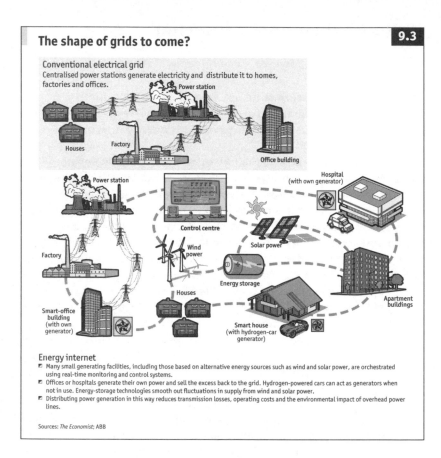

Conventional electrical grid
Centralised power stations generate electricity and distribute it to homes, factories and offices.

Power station

Houses

Factory

Office building

Power station

Control centre

Wind power

Solar power

Energy storage

Houses

Factory

Smart-office building (with own generator)

Smart house (with hydrogen-car generator)

Hospital (with own generator)

Apartment buildings

Energy internet

- Many small generating facilities, including those based on alternative energy sources such as wind and solar power, are orchestrated using real-time monitoring and control systems.
- Offices or hospitals generate their own power and sell the excess back to the grid. Hydrogen-powered cars can act as generators when not in use. Energy-storage technologies smooth out fluctuations in supply from wind and solar power.
- Distributing power generation in this way reduces transmission losses, operating costs and the environmental impact of overhead power lines.

Sources: *The Economist*; ABB

local utility's managers had to ask the regional system operator by phone what was happening on their own wires. Meanwhile, the failure cascaded to neighbouring regions. "They simply can't see the grid!" laments Clark Gelling of the EPRI.

Even if operators had smart sensors throughout the system, they could do little to halt problems from spreading, because they lack suitable control systems. Instead, essential bits of energy infrastructure are built to shut down at the first sign of trouble, spreading blackouts and increasing their economic impact. The North American blackout, for example, cost power users around $7 billion. Engineers have to spend hours or even days restarting power plants.

The good news is that technologies are now being developed in four areas that point the way towards the smart grid of the future. First, utilities are experimenting with ways to measure the behaviour of the grid in real time. Second, they are looking for ways to use that information to control the flow of power fast enough to avoid blackouts. Third, they are upgrading their networks in order to pump more juice through the grid safely. Last, they are looking for ways to produce and store power close to consumers, to reduce the need to send so much power down those ageing transmission lines in the first place.

First, to the eyes and ears. With the exception of some simple sensors located at a minority of substations, there is little "intelligence" embedded in today's grid. But in America's Pacific north-west, the Bonneville Power Administration (BPA), a regional utility run by the federal government, has been experimenting with a wide-area monitoring system. Carson Taylor, BPA's chief transmission expert, explains that the impetus for this experiment was a big blackout in 1996. Sensors installed throughout the network send data about local grid conditions to a central computer, 30 times a second.

Dr Taylor credits this system with preventing another big blackout in his region, and says his counterparts in America's north-east could have avoided 2003's blackout if they had had such a system. He wishes his neighbours to the south, in power-starved California, who import hydroelectric power from Canada over BPA's transmission lines, would upgrade their networks too. If they did, he believes the entire western region could enjoy a more reliable power supply.

Real-time data is, of course, useless without the brains to process it and the brawn to act on it. For the brains, look to Roger Anderson and his colleagues at Columbia University and at the Texas Energy Centre. They are developing software to help grid managers make sense of all that real-time data, and even to forecast problems before they occur. They hope to use the Texas grid, which (fittingly, for the Lone Star state) stands alone from North America's eastern and western power grids, as a crucible for their reforms. ABB, a Swiss-Swedish engineering giant, has also developed brainy software that tracks grid flows several times a second and feeds the information to control systems that can respond within a minute or so. The firm claims it can make outages 100 times less likely. The real challenge is responding in real time. Today's electromechanical switches take tenths of seconds or longer to divert power – usually far too long to avoid a problem. But several firms have devised systems that can switch power in milliseconds. At the Marcy substation in upstate New York, the New York Power Authority and the EPRI are experimenting with a device that can instantaneously switch power between two transmission lines – one notoriously congested, the other usually not – that bring power into New York City.

Another bit of brawn comes in the shape of devices that can act as "shock absorbers" and smooth out fluctuations in the power supply. Greg Yurek, the head of American Superconductor and a former professor at the Massachusetts Institute of Technology, argues that recent trends have increased the instability of the grid and highlighted the need for this sort of technology. In America, deregulation of the wholesale power market means ever larger quantities of power are travelling greater distances, yet investment in the grid has halved since the 1970s.

Traditionally, grid operators used banks of capacitors, which store and release energy, to act as shock absorbers for the grid. But capacitor banks tend to be very large and hard to site near customers (who love to guzzle power but complain about new power lines or hardware in their neighbourhood). American Superconductor makes smarter devices known as D-VARs that fit into portable tractor-trailers and can be parked right next to existing substations. Clever software monitors the grid and responds in a matter of milliseconds if it detects fluctuations.

The third broad area of improvement involves squeezing more juice through existing power lines. It may not be necessary to lay thousands of miles of new copper cables to tackle this problem. Because of the current lack of real-time monitoring and controls, system operators often insist that utilities run just 50% of the maximum load through their wires. That safety margin is probably prudent today. But as the grid gets smarter in various ways, EPRI officials reckon that it may be possible to squeeze perhaps a third more juice through today's wires.

And if those copper wires were replaced with something better, even more power could be piped through the grid. One alternative is a cable that uses a combination of aluminium and carbon-glass fibre composite. Researchers at CTC, a cable-maker working with the University of Southern California, think this composite cable could carry twice as much power as a conventional one. Similarly, American Superconductor has come up with superconducting cables that can carry five times as much power as ordinary wires.

Back to the future

In the long run, however, the solution surely does not lie in building ever fatter pipes to supply ever more power from central power plants to distant consumers. Amory Lovins, head of the Rocky Mountain Institute, an environmental think-tank, explains why: "the more and bigger bulk power lines you build, the more and bigger blackouts are likely." A better answer is "micropower" – a large number of small power sources located near to end-users, rather than a small number of large sources located far away.

This sentiment is echoed by experts at America's Carnegie Mellon and Columbia universities, who have modelled the vulnerabilities (to trees or terrorists) of today's brittle power grid. Even the gurus at EPRI, which relies on funding from utilities that run big power plants, agree that moving to a distributed model, in conjunction with a smarter grid, will reduce blackouts. Look at Denmark, which gets around 20% of its power from scattered wind farms, for example. Sceptics argued that its reliance on micropower would cause more blackouts. It did not.

At first glance, this shift toward micropower may seem like a return to electricity's roots over a century ago. Thomas Edison's original vision was to place many small power plants close to consumers. However, a complete return to that model would be folly, for it would rob both the grid and micropower plants of the chance to sell power when the other is in distress. Rather, the grid will be transformed into a digital network capable of handling complex, multi-directional flows of power. Micropower and megapower will then work together.

ABB foresees the emergence of "microgrids" made up of all sorts of distributed generators, including fuel cells (which combine hydrogen and oxygen to produce electricity cleanly), wind and solar power. The University of California at Irvine is developing one, as are some firms in Germany. "Virtual utilities" would then aggregate the micropower from various sources in real time – and sell it to the grid.

Energy-storage devices will be increasingly important too. Electricity, almost uniquely among commodities, cannot be stored efficiently (except as water in hydroelectric dams). That means grid operators must match supply and demand at all times to prevent blackouts. But if energy could be widely stored on the grid in a distributed fashion, and released cheaply and efficiently when needed, it would transform the reliability and security of the grid. According to Dr Schainker, the last few years have brought dramatic advances in this area. He reckons that several energy-storage technologies now look quite promising: advanced batteries, flywheels and superconducting devices known as SMES devices. But the most intriguing storage option involves hydrogen – which can be used as a medium to store energy from many different sources.

Most of the hoopla surrounding hydrogen has concentrated on its role in powering fuel-cell cars. However, its most dramatic impact may well come in power generation. That is because hydrogen could radically alter the economics of intermittent sources of green power. At the moment, much wind power is wasted because the wind blows when the grid does not need, or cannot safely take, all that power. If that wasted energy were instead stored as hydrogen (produced by using the electrical power to extract hydrogen from water), it could later be converted back to electricity in a fuel cell, to be sold when needed. Geoffrey Ballard of Canada's General Hydrogen, and the former head of Ballard, a leading fuel-cell-maker, sees hydrogen and electricity as so interchangeable on the power grid of the future that he calls them "hydricity".

Another benefit is that hydrogen could be sold to allow passing fuel-cell-powered electric cars to refill their tanks. In time, those automobiles might themselves be plugged into the grid. Tim Vail of General Motors calculates that the power-generation capacity trapped under the hoods of the new cars sold in America each year is greater than all the country's nuclear, coal and gas power plants combined. Most cars are in use barely a tenth of the time. If even a few of them were plugged into the grid (in a car park, say), a "virtual utility" could tap their generating power, getting them to convert hydrogen into electricity and selling it on to the grid for a tidy profit during peak hours, when the grid approaches overload.

Brighter prospects?

So, given all of the environmental, economic and energy benefits of upgrading the power grid, will it really happen? Do not hold your breath. The EPRI reckons that building an energy internet could cost over $200 billion in America alone. Even so, the obstacle to progress, in America at least, is not really money. For even $200 billion is not an outrageous amount of money when spread over 20 or 30 years by an industry with revenues of over $250 billion.

The snag is politics: America's half-baked attempt at deregulation has drained the industry of all incentives for grid investment. America's power industry reinvests less than 1% of its turnover in research and development – less than any other big industry. While Britain is a notable exception, the picture is not much better in many parts of the world. The technology exists to enable a radical overhaul of the way in which energy is generated, distributed and consumed – an overhaul whose impact on the energy industry could match the internet's impact on communications. But unless regulators restore the economic incentives for investment, the future looks bleak. Time to stock up on candles and torches.

The material on pages 285–90 first appeared in The Economist in March 2004.

Why the future is hybrid

Hybrid petrol-electric cars such as the Toyota Prius are becoming increasingly popular. But are they any more than a rest-stop on the road to the hydrogen car?

W HY HAS THE TOYOTA PRIUS become the car industry's most talked about product? Since 1997, only about 250,000 have been sold, a paltry number by the industry's standards. The Prius is hardly big, fast or beautiful – the attributes that usually appeal to commentators, aficionados or, for that matter, buyers. And yet it is significant because it is the world's first mass-produced petrol-electric hybrid car, powered by both an internal-combustion engine and an electric motor. The second-generation Prius, launched in 2003, won some of the industry's most prestigious awards – it was named European Car of the Year 2005 – and generated a buzz out of all proportion to the car's prevalence on the roads.

By choosing to drive a Prius, buyers can demonstrate how green they are without paying any penalty other than a slightly higher purchase price. Compared with a new American car of the same size, the Prius consumes roughly half as much petrol, and so releases half as much climate-changing carbon dioxide. Moreover, its emissions of smog-forming pollutants, such as nitrogen oxides and hydrocarbons, are 90% lower. Yet the Prius still manages to deliver the comfort and performance of a conventional car.

The success of the Prius has taken Toyota by surprise. The average wait at American dealerships in December 2004 was six months, even though the company increased its sales target for North America from its initial estimate of 36,000 units to 47,000 for 2004. To meet demand, Toyota announced another increase, saying it would push monthly global production up in 2005 by 50% to 15,000 cars, and double its allotment for America to 100,000 units. While that number is still only one-quarter of 2003's sales for America's most popular Toyota model, the Camry, it shows that consumers are willing to pay a premium for clean, environmentally friendly cars – as long as there is no need to compromise on performance.

Other carmakers are scurrying to catch up. CSM Worldwide, an automotive research firm, reckons that at least 20 new hybrid models will appear in America by 2007. Besides 2004's new Ford Escape and Honda Accord hybrids, Toyota added two sport-utility vehicles (SUVS) to its hybrid line-up early in 2005. DaimlerChrysler announced that it would introduce a Mercedes hybrid by 2009, and Porsche is considering a hybrid version of its Cayenne SUV. Even General Motors, one of the strongest proponents of hydrogen fuel-cell cars, has jumped on the hybrid bandwagon with two pick-up trucks, a sedan and several SUVS to follow. Thanks to the convergence of geopolitics, technology and fashion, hybrids are picking up speed.

An old new idea

While the arrival of mass-produced hybrids is new, the idea itself is not. Indeed, it dates back to early automotive history when cars powered by electric motors, steam or internal-combustion engines all accounted for significant shares of the market. Why hybrids failed then is best illustrated by the example of an American engineer named H. Piper, who filed a patent for a petrol-electric hybrid vehicle in 1905. His idea was to use an electric motor to assist an internal-combustion engine, enabling it to achieve a thrilling 40kph (25mph). Unfortunately for Mr Piper, petrol-powered internal-combustion engines achieved those speeds on their own just a few years later, undermining the more complex and expensive hybrid approach. Petrol engines soon ruled the roost.

Priorities began to change in the early 1970s, when the oil crisis increased demand for less fuel-thirsty cars. As a result, the overall fuel efficiency of cars and trucks improved dramatically (though it stalled in America in the late 1980s as cheap petrol and a regulatory loophole encouraged sales of SUVS and light trucks). Moreover, in the 1990s, concern began to grow over the impact of fossil-fuel consumption on climate change.

During the 1990s, all of the big three American carmakers developed diesel-electric hybrid concept cars, though none made it into production. Instead, the focus shifted to pure-electric vehicles, which are technologically simpler than hybrids. But their high cost and limited range deterred consumers. Even the most advanced models could only go about 100 miles before they needed to be plugged in and recharged for several hours. By 2000, most electric cars had been taken out of production.

Meanwhile, Toyota released its first Earth Charter in 1992, setting the goal of minimising its overall environmental impact. In September 1993, the company began to plan the development of a car for the next century, dubbed Globe 21st Century, or G21. Originally, the plan was to produce a car with 50% better fuel economy than existing vehicles. But over the course of the project this target was raised to 100%, at which point it became clear that tweaking a petrol engine would not suffice. Instead, a more radical solution would be needed: a hybrid.

Despite the higher cost and complexity of a hybrid system, Toyota decided to press ahead with a massive research and development effort. Improved technology – such as better batteries and cheaper, more powerful control electronics to co-ordinate the two propulsion systems – meant that a mass-produced hybrid was now feasible. In 1997, the Prius was launched in Japan. It was followed by Honda's Insight hybrid in 1999.

When the Prius went on sale in America in 2000, it did not cause much of a stir. Indeed, in 2003, Honda and Toyota sold about the same number of hybrids in America. In 2004, however, Toyota sold about twice as many as Honda. The Prius took off thanks to the combination of rising petrol prices, celebrity endorsements and a futuristic redesign. (There is no petrol version of the Prius, so the car makes a statement in a way that the Honda Civic, which is available in both petrol and hybrid versions, does not.) It is the first hybrid to become a hit.

Hybrid anatomy

There is more to the Prius than clever marketing, however. To understand why, it is necessary to look under the bonnet at the way different kinds of hybrids work – for not all hybrids are the same. The simplest kind is the "stop-start" or "micro" hybrid, which is not generally regarded as a true hybrid because it relies solely on an internal-combustion engine for propulsion. As the "stop-start" name implies, the engine shuts off when the vehicle comes to a halt. An integrated starter-generator restarts the engine instantly when the driver steps on the accelerator. All of this increases fuel efficiency only slightly, typically by around 10%. But few modifications to a conventional design are required, so it costs very little. In Europe, PSA Peugeot Citroën introduced a stop-start version of the Citroën C3, which sells for roughly the same price as a similarly equipped conventional C3.

Next come so-called "mild" hybrid designs, such as Honda's Integrated Motor Assist (IMA) – the hybrid configuration found in the Insight, the Civic and the Accord. In addition to a stop-start function, an electric motor gives the engine a boost during acceleration. During braking, the same motor doubles up as a generator, capturing energy that

9.4

Hybrid vigour
How the Prius works

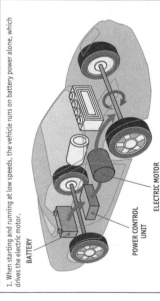

1. When starting and running at low speeds, the vehicle runs on battery power alone, which drives the electric motor.

BATTERY

POWER CONTROL UNIT

ELECTRIC MOTOR

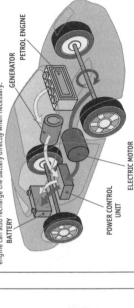

2. In normal driving conditions, power from the petrol engine is divided and used both to drive the wheels directly, and to turn the generator, which in turn drives the electric motor.

GENERATOR

PETROL ENGINE

POWER CONTROL UNIT

ELECTRIC MOTOR

3. When sudden acceleration is needed, the battery provides extra power to the electric motor, supplementing the power from the petrol engine.

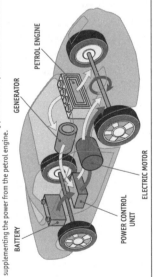

PETROL ENGINE

GENERATOR

BATTERY

POWER CONTROL UNIT

ELECTRIC MOTOR

4. The battery is recharged in two ways. When braking, the electric motor acts as a generator, converting the vehicle's kinetic energy into electrical energy and storing it in the battery. The engine can also recharge the battery directly when necessary.

GENERATOR

PETROL ENGINE

BATTERY

POWER CONTROL UNIT

ELECTRIC MOTOR

Source: Toyota

294

would otherwise be lost as heat and using it to recharge the car's batteries. Since the electric motor is coupled to the engine, it never drives the wheels by itself. That is why this system is called a mild hybrid, much to Honda's dismay. The design is less expensive than Toyota's more elaborate approach, but can provide many of the same benefits, says Dan Benjamin of ABI Research, a consultancy based in Oyster Bay, New York. The hybrid version of the Civic achieves 48 miles per gallon, a 37% improvement over a comparable conventional Civic.

Toyota's Hybrid Synergy Drive, a "full" hybrid system, is much more complex. (The Ford Escape hybrid uses a similar system; Ford licenses a number of patents from Toyota.) Using a "power split" device, the output from the petrol engine is divided and used both to drive the wheels directly and to turn the generator, which in turn drives the electric motor and also drives the wheels. The distribution of power is continuously variable, explains David Hermance of Toyota, allowing the engine to run efficiently at all times. When its full power is not needed to drive the wheels, it can spin the generator to recharge the batteries. The batteries also get replenished when the car is coasting or braking. During stop-and-go traffic and at low speeds, when the petrol engine would be most inefficient, it shuts off and the electric motor, powered by the battery, takes over. That explains why the Prius has a better fuel economy rating for urban driving (60 miles per gallon) than for motorway driving (51 miles per gallon): the opposite of a conventional vehicle.

The next step may be the "plug-in" hybrid, which is not the backwards step its name suggests. Unlike the electric cars of the 1990s, none of today's hybrids needs to be plugged in – but if plugging were an option it would be a good idea. Andrew Frank and his team at the University of California Davis' Hybrid Electric Vehicle Centre are working exclusively on plug-in hybrids, which can operate as pure-electric vehicles over short distances (up to 60 miles, with a large enough battery pack) but can switch to a hybrid system when needed. Since the average American driver travels about 30 miles a day, plug-in hybrids could be recharged overnight, when electricity is cheaper to produce, and need never use petrol at all, except on longer trips.

According to studies carried out by the Electric Power Research Institute (EPRI), plug-in hybrids could be one of the cleanest and most efficient kinds of car. In 2002, the EPRI teamed up with DaimlerChrysler to build five plug-in hybrid vans, the first of which was unveiled at a trade show in September. The larger battery packs make the upfront costs for plug-ins higher than for other hybrids. But Bob Graham of the EPRI says

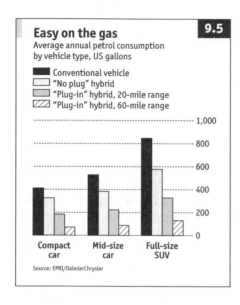

Easy on the gas `9.5`

Average annual petrol consumption by vehicle type, US gallons

- Conventional vehicle
- "No plug" hybrid
- "Plug-in" hybrid, 20-mile range
- "Plug-in" hybrid, 60-mile range

Compact car Mid-size car Full-size SUV

Source: EPRI/DaimlerChrysler

the added costs could be more than recouped over the vehicle's life.

Not everyone is bothered by high fuel consumption, however, as the current enthusiasm for enormous SUVs demonstrates. So hybrids seem likely to remain a niche: ABI Research predicts that by 2010, less than 5% of all cars sold in America will be hybrids, assuming current petrol prices persist. But if Alan Lloyd has his way, hybrids and other low-emission vehicles will become far more commonplace. Dr Lloyd is head of the California Air Resources Board (CARB), a state agency that enforces arguably the most stringent air-quality rules in the world. California has passed landmark legislation to curb the emissions of greenhouse gases by 30% beginning in 2009. Since carbon-dioxide emissions are directly linked to a car's fuel consumption, critics charge that the new rules are in effect a way to legislate fuel economy, which is supposed to be regulated by the federal government, not the states. As a result, carmakers are expected to challenge the new rules in court.

Sales of hybrids in Europe are a fraction of those in America. Instead, diesel cars have become Europe's answer to reduce fuel consumption, curb greenhouse emissions and save money at the pump. Because diesel fuel contains more energy per unit, the fuel economy of diesel cars is roughly 30% better than that of petrol-powered cars. Moreover, diesel cars are not as loud or dirty as they once were, thanks to technologies such as electronically controlled "common rail" fuel-injection systems. Diesels now make up about 45% of all newly registered cars in Europe.

Even so, they still lag behind petrol engines in terms of cleanliness. In the process of combustion, diesels create a lot of pollution, including nitrogen oxides which cause smog, and particulate matter that can cause respiratory problems. That said, some carmakers have begun to equip their cars with particulate filters, notably PSA Peugeot Citroën. Together with two British firms, Ricardo and QinetiQ, the company is building a diesel-hybrid based on the family-sized Citroën Berlingo. The aim is to achieve a combined fuel economy of 70 miles per gallon with carbon-dioxide emissions of only 90 grams per kilometre. (In comparison, the Prius delivers 55 miles per gallon with carbon-dioxide emissions of 104 grams per kilometre.)

While it is uncertain whether the car will be mass produced, it is clear that a diesel-electric hybrid would make for an extremely frugal vehicle. A study by the Laboratory for Energy and the Environment at the Massachusetts Institute of Technology, which looked at energy use over the course of a vehicle's life, predicts that by 2020, diesel hybrids could achieve the same energy-efficiency and greenhouse-gas emissions as fuel-cell cars powered by hydrogen made from natural gas. The difference is that diesel-hybrid technology is available today.

So why are diesel hybrids taking so long to appear on the roads? Hybrid diesels impose a double price premium, explains Lindsay Brooke, an analyst at CSM Worldwide. Combining a diesel engine, (which costs around $2,000 more than a petrol engine) with a hybrid powertrain (which adds another $3,000 or so) would make for an expensive proposition. Systems to treat the exhaust would impose further costs. The prospects for diesels and diesel hybrids are particularly dim in America, where regulations in California (and, from 2007, nationwide) require diesels to be as clean as petrol-driven cars. Some progress has been made: particulate filters can now eliminate more than 90% of diesel soot. But traps for nitrogen oxides remain a challenge.

The car of the future, today

Hydrogen fuel-cell vehicles promise to be the cleanest mode of transportation, eliminating harmful tailpipe emissions altogether. But despite much publicity, and the fact that most carmakers are working on the technology, fuel-cell cars will not appear in significant quantities any time soon. America's National Academy of Sciences, which advises the government on new technologies, recently estimated that the transition to a "hydrogen economy" will probably take decades, since many challenges remain – in particular, how to produce, store and distribute hydrogen in sufficient quantities.

Hybrid cars, however, offer many of the benefits of fuel-cell vehicles, with the huge advantage that they are available now. Moreover, as the success of the Prius shows, people will actually buy them. The beauty of petrol-electric hybrids is that they do not require any changes in driver behaviour or the fuel-delivery infrastructure.

Rather than being mere stepping-stones on the way to the hydrogen cars of the future, petrol-electric hybrids are likely to be around for years, if not decades, to come. When and if fuel-cell cars become available down the road, they may not replace hybrids, but instead are likely to be descended from them, since they require many of the same components, from control systems to motors. As Joseph Romm, director of the Centre for Energy & Climate Solutions, a non-profit organisation based in Arlington, Virginia, puts it, "hybrids are almost certainly the platform from which all future clean vehicles will evolve."

The material on pages 291–8 first appeared in *The Economist* in December 2004.

The rise of the green building

New buildings use design and technology to reduce environmental impact, cut costs and provide better places to work

IT IS OFFICIALLY KNOWN AS THE SWISS RE TOWER, or 30 St Mary Axe. But Londoners universally refer to the newest addition to their skyline as "the Gherkin", thanks to the 41-storey building's distinctive, curved profile, which actually looks more like a pine cone. What is most remarkable about the building is not its name or its shape, however, but its energy-efficiency. Thanks to its artful design and some fancy technology, it is expected to consume up to 50% less energy than a comparable conventional office building.

Most people are not used to thinking of large buildings as vast, energy-guzzling machines. But that is what they are. In America, buildings account for 65% of electricity consumption, 36% of total energy use and 30% of greenhouse-gas emissions. So making buildings more energy-efficient could have a significant impact on energy policy, notes Rebecca Flora of the Green Building Alliance, a group that promotes sustainable architecture. That is a key goal of the "green architecture" movement, which is changing the way buildings are designed, built and run.

Proponents of green architecture argue that the approach has many benefits. In the case of a large office, for example, the combination of green design techniques and clever technology can not only reduce energy consumption and environmental impact, but also reduce running costs, create a more pleasant working environment, improve employees' health and productivity, reduce legal liability, and boost property values and rental returns.

The term "green architecture" only came into use in the 1990s, but the movement's roots can be traced back a long way. Crystal Palace in London and Milan's Galleria Vittorio Emanuele II, for example, built in 1851 and 1877 respectively, used roof ventilators and underground air-cooling chambers to regulate the indoor temperature. Today's enthusiasm for green architecture has its origins in the energy crisis of the 1970s, when architects began to question the wisdom of building enclosed glass-and-steel boxes that required massive heating and cooling systems. Early proponents of more energy-efficient architecture included

William McDonough, Bruce Fowle and Robert Fox in America, Thomas Herzog in Germany, and Norman Foster and Richard Rogers in Britain.

These forward-thinking architects began to explore designs that focused on the long-term environmental impact of maintaining and operating a building, looking beyond the so-called "first costs" of getting it built in the first place. This approach has since been formalised in a number of assessment and rating systems, such as the BREEAM standard introduced in Britain in 1990, and the LEED (Leadership in Energy and Environmental Design) standards developed by the United States Green Building Council (USGBC) starting in 2000.

The LEED standards are intended to produce "the world's greenest and best buildings" by giving developers a straightforward checklist of criteria by which the greenness of a building can be judged. Points are awarded in various categories, from energy use (up to 17 points) to water-efficiency (up to five points) to indoor environment quality (up to 15 points); the total then determines the building's LEED rating. Extra points can be earned by installing particular features, such as renewable-energy generators or carbon-dioxide monitoring systems. A building that achieves a score of 39 points earns a "gold" rating; 52 points earns a "platinum" rating. A gold-rated building is estimated to have reduced its environmental impact by 50% compared with an equivalent conventional building, and a platinum-rated building by over 70%.

Rating buildings in this way reveals how inefficient traditional buildings and building processes are. "We can sometimes waste up to 30 cents on the dollar," says Phillip Bernstein, an architect and professor at Yale University. "It's not just the consumption of energy, it's the use of materials, the waste of water, the incredibly inefficient strategies we use for choosing the subsystems of our buildings. It's a scary thing." In part, he says, this is because the construction industry is so fragmented. Designers, architects, engineers, developers and builders each make decisions that serve their own interests, but create huge inefficiencies overall.

Green is good

But things are now changing, as green architecture moves into the mainstream. In the spring of 2003, Toyota completed a 624,000-square-foot office complex in Torrance, California, that received a LEED gold rating, thanks to the inclusion of features such as solar cells to provide up to 20% of the building's energy needs. Also in 2003, Pittsburgh opened the doors on its 1.5m-square-foot convention centre, the largest building to be awarded a gold LEED rating so far. The USGBC says nearly 1,700 buildings in 50 states are seeking LEED certification and 137 had been constructed and certified by the end of 2004. And America's General Services Administration, which oversees all non-military government construction, has decreed that all new projects and renovations must meet the minimum LEED standards.

In Britain, meanwhile, 70 office buildings constructed during 2003, representing 25% of the total by floor area, met the BREEAM standard. Similar standards have been adopted in New Zealand, Australia and Canada. In China, the Beijing Organising Committee of the Olympic Games aims to host the first zero-net-emissions games, which will include constructing all buildings and sports venues using green-architecture principles.

There are many ways to reduce a building's environmental impact. Consider the 48-storey Condé Nast Building at 4 Times Square in New York, for example, which was designed by Fox & Fowle Architects. It was one of the first examples in which green-architecture principles were applied to a large urban office building, and informed the drawing up of the LEED points system, since it uses almost every energy-saving technique imaginable.

Special glass allows daylight in to reduce the need for interior lighting, keeps heat and ultraviolet rays out, and minimises heat loss in winter. Two natural-gas-powered fuel cells provide 400 kilowatts of power, enough to provide all the electricity needed at night, and 5% of the building's needs during the day. The hot-water exhaust produced by the fuel cells is used to help heat the building and provide hot water. The heating and cooling systems, located on the roof, are gas-powered rather than electric, which reduces energy losses associated with electrical power transmission. Photovoltaic panels on the building's exterior provide up to an additional 15 kilowatts of power. Inside the building, motion sensors control fans and switch off lights in seldom-occupied areas such as stairwells. Exit signs are illuminated by low-power light-emitting diodes. The result is that the building's energy consumption is 35–40% lower than that of a comparable conventional building.

30 St Mary Axe, designed by Foster and Partners, is also packed with energy-saving features. In particular, it uses natural lighting and ventilation wherever possible. The façade consists of two layers of glass (the outer one double-glazed) enclosing a ventilated cavity with computer-controlled blinds. A system of weather sensors on the outside of the building monitors the temperature, wind speed and level of sunlight,

closing blinds and opening window panels as necessary. The building's shape maximises the use of natural daylight, reducing the need for artificial lighting and providing impressive long-distance views even from deep inside the building.

The highest-profile green building currently on the drawing board is the Freedom Tower, which will be built on the site of the World Trade Centre in New York. The architects, Skidmore, Owings & Merrill and Studio Daniel Libeskind, have incorporated environmental design features throughout the huge complex. The main tower, which will rise 1,776 feet, will include solar panels and a wind farm, the turbines of which are expected to deliver around one megawatt of power, enough to provide up to 20% of the building's expected demand. Like other green buildings, it will rely on natural light and ventilation, and energy-efficient lighting.

High energy costs, environmental concerns and anxiety about the "sick building syndrome" associated with the sealed-box structures of the 1970s all helped to jump-start the green-architecture movement. But now economics is driving the shift towards greener design, as new materials and techniques fall in price, argues Michael Crosbie, an architect at Steven Winter Associates, a consultancy based in Norwalk, Connecticut. He says his clients "are much more demanding because they see the incredible amount of money it takes to get something constructed, and they want a return on that investment."

Why it pays to be green

Going green saves money by reducing long-term energy costs: a survey of 99 green buildings in America found that on average, they use 30% less energy than comparable conventional buildings. So any additional building costs can be recovered quickly: according to the USGBC, the 2% increase in construction costs required to achieve a LEED gold rating typically pays for itself in lower running costs within two years. The traditional approach of trying to minimise construction costs, by contrast, can lead to higher energy bills and wasted materials.

Energy-saving techniques need not all be as exotic as installing coated glass, computer-controlled blinds or photovoltaic cells. Mr Crosbie says builders are now insulating buildings more effectively, in some cases using materials such as recycled paper and fabrics, including old, shredded jeans. It is more effective than traditional insulation, he says, saves money and is easier on the environment.

Green buildings can also have less obvious economic benefits. The use of natural daylight in office buildings, for example, as well as reducing energy costs, seems to make workers more productive. Studies conducted by Rachel and Stephen Kaplan, environmental psychologists at the University of Michigan, found that employees with views of a natural landscape report greater job satisfaction, less stress and fewer illnesses. Lockheed Martin, an aerospace firm, found that absenteeism fell by 15% after it moved 2,500 employees into a new green building in Sunnyvale, California. The increase in productivity paid for the building's higher construction costs within a year.

Similarly, the use of daylight in shopping complexes appears to increase sales. The Heschong Mahone Group, a California-based consultancy that specialises in energy-efficient building technologies, found that sales were as much as 40% higher in stores lit with skylights. It also found that students in naturally lit classrooms performed up to 20% better. Green buildings can also reduce legal liabilities for their owners, since they are less likely to give rise to "sick building" lawsuits. But more studies are needed, says Caren Glotfelty, director of the environmental programme at the Heinz Endowments, a non-profit foundation run by Teresa Heinz Kerry that funds sustainable initiatives.

Despite its benefits and its growing popularity, green architecture is still the exception, not the rule, however. The main problem is co-ordination, says Mr Bernstein, who is also vice-president of the building solutions division at Autodesk, a software company. Green buildings require much more planning by architects, engineers, builders and developers than traditional buildings. "The building industry is very disaggregated," he says, "so adoption patterns are really, really slow." But new software is now improving planning by simulating how a building will perform before it is built.

Autodesk's software can create a three-dimensional model of a building and then work out how much energy it will use, taking into account its shape, heating and cooling systems, orientation to the sun and geographic location. Other such tools abound: the designers of 4 Times Square calculated its energy consumption using a free package called DOE-2, developed by James J. Hirsch & Associates together with the Lawrence Berkeley National Laboratory, with funding from America's Department of Energy.

Greener by design

In the old days, says Mr Bernstein, assessing a building's environmental impact had to be done with spreadsheets, calculators and informed guessing, and three-dimensional modelling was primarily used to prepare presentations. But now the three-dimensional computer models are being used with sophisticated analytical tools. "We are getting to the next phase where you can analyse rather than simply represent," he says. It is then possible to predict how much energy and water a building will consume, how much material will be needed, and other parameters that determine its LEED certification. All of this is old hat for the airline and automobile industries, where computer models have long been used to trim costs and streamline design before construction begins. Now the same technology is being applied by architects.

Computers also make possible entirely new designs. 30 St Mary Axe, for example, could not have been built without a computer model to specify the exact shape of every one of its 5,500 glass panels, or to model the airflow in and around it. Similarly, computer modelling made possible the Avax office building completed in Athens, Greece, in 1998. It has sheaves of glass which open and close automatically, depending on the intensity and angle of the sun, to provide sunlight while preventing the building from overheating. The ventilation system in Pittsburgh's convention centre uses the natural "chimney effect" created by its sweeping roof to draw air through vents by the river below, cooling the building without using a single fan.

This is more than a mere fad, or the use of technology for the sake of it, says Mr Bernstein. Green architecture will, he suggests, help to reshape the construction industry over the next few years, with ever more innovative, energy-efficient and environmentally friendly buildings. "No one is doing this for fun," he says. "There's too much at stake."

The material on pages 299–304 first appeared in *The Economist* in December 2004.

10
SMALL WONDERS

Small wonders

Nanotechnology will give humans greater control of matter at tiny scales. That is a good thing

ATOMS ARE THE FUNDAMENTAL BUILDING BLOCKS of matter, which means they are very small indeed. The world at the scale of atoms and molecules is difficult to describe and hard to imagine. It is so odd that it even has its own special branch of physics, called quantum mechanics, to explain the strange things that happen there. If you were to throw a tennis ball against a brick wall, you might be surprised if the ball passed cleanly through the wall and sailed out on the other side. Yet this is the kind of thing that happens at the quantum scale. At very small scales, the properties of a material, such as colour, magnetism and the ability to conduct electricity, also change in unexpected ways.

It is not possible to "see" the atomic world in the normal sense of the word, because its features are smaller than the wavelength of visible light (see Chart 10.1). But back in 1981, researchers at IBM designed a probe called the scanning tunnelling microscope (STM), named after a quantum-mechanical effect it employs. Rather like the stylus on an old-fashioned record player, it could trace the bumps and grooves of the nanoscale world. This allowed scientists to "see" atoms and molecules for the first time. It revealed landscapes as beautiful and complex as the ridges, troughs and valleys of a Peruvian mountainside, but at the almost unimaginably small nanometre (nm) scale.

A nanometre is a billionth of a metre, or roughly the length of ten hydrogen atoms. Although scientists had thought about tinkering with things this small as long ago as the late 1950s, they had to wait until the invention of the STM to make it possible.

Nanotechnology is generally agreed to cover objects measuring from 1 to 100nm, though the definition is somewhat arbitrary. Some people include things as small as one-tenth of a nanometre, which is about the size of the bond between two carbon atoms. At the other end of the range, in objects larger than 50nm the laws of classical physics become increasingly dominant.

There are plenty of materials that simply happen to have features at the nanoscale – such as stained glass, mayonnaise or cat litter – but do not qualify for the nanotechnology label. The point about nanotechnology is that it sets out deliberately to exploit the strange properties found in these very small worlds.

At the nanoscale, explains George Smith, the amiable head of materials science at Oxford University, "new, exciting and different" properties can be found. If you were to start with a grain of sugar, he says, and chopped it up into ever smaller pieces and simply ended up with a tiny grain of sugar, that would be no big deal. But as an

From ants to atoms	10.1

1 millimetre = 1m nanometres (nm)

Item	Size in nm
Red ant	5m
Human hair (width)	80,000
Diameter of a typical bacterium	1,000–10,000
Average wavelength of visible light	400–700
Human immunodeficiency virus	90
Wavelength of extreme ultraviolet light	40
Cell membrane	10
Diameter of DNA	~2.5
Ten hydrogen atoms	1
Water molecule (width)	0.3

Sources: Wikipedia; National Institute of Standards and Technology; Intel; Royal Society; R. Smalley

object gets smaller, the ratio between its surface area and its volume rises. This matters because the atoms on the surface of a material are generally more reactive than those at its centre.

So icing sugar, for instance, dissolves more quickly in water than does the granulated form. And if silver is turned into very small particles, it has antimicrobial properties that are not present in the bulk material. One company exploits this phenomenon by making nanoparticles of the compound cerium oxide, which in that form are chemically reactive enough to serve as a catalyst.

In this invisible world, tiny particles of gold melt at temperatures several hundred degrees lower than a large nugget, and copper, which is normally a good conductor of electricity, can become resistant in thin layers in the presence of a magnetic field. Electrons, like that imaginary tennis ball, can simply jump (or tunnel) from one place to another, and molecules can attract each other at moderate distances. This effect allows geckos to walk on the ceiling, using tiny hairs on the soles of their feet.

But finding novel properties at the nanoscale is only the first step. The next is to make use of this knowledge. Most usefully, the ability to make stuff with atomic precision will allow scientists to produce materials with improved, or new, optical, magnetic, thermal or electrical properties. And even just understanding the atomic-scale defects in a material can suggest better ways of making it.

Indeed, entirely new kinds of material are now being developed. For example, NanoSonic in Blacksburg, Virginia, has created metallic rubber, which flexes and stretches like rubber but conducts electricity like a solid metal. General Electric's research centre in Schenectady in New York state is trying to make flexible ceramics. If it succeeds, the material could be used for jet-engine parts, allowing them to run at higher, more efficient temperatures. And several companies are working on materials that could one day be made into solar cells in the form of paint.

Because nanotechnology has such broad applications, many people think that it may turn out to be as important as electricity or plastic. As this section will show, nanotechnology will indeed affect every industry through improvements to existing materials and products, as well as allowing the creation of entirely new materials. Moreover, work at the smallest of scales will produce important advances in areas such as electronics, energy and biomedicine.

From small beginnings

Nanotechnology does not derive from a single scientific discipline. Although it probably has most in common with materials science, the properties of atoms and molecules underpin many areas of science, so the field attracts scientists of different disciplines. Worldwide, around 20,000 people are estimated to be working in nanotechnology, but the sector is hard to define. Small-scale work in electronics, optics and biotechnology may have been relabelled "nanobiotechnology", "nano-optics" and "nanoelectronics" because nano-anything has become fashionable.

The "nano" prefix is thought to derive from the Greek noun for dwarf. Oxford's Mr Smith jokingly offers an alternative explanation: that it "comes from the verb which means to seek research funding". And research funding is certainly available by the bucketload. Lux Research, a nanotechnology consultancy based in New York, estimates that total spending on nanotechnology research and development by governments, companies and venture capitalists worldwide was more than $8.6 billion in 2004, with over half coming from governments. But Lux predicts that in future years companies are likely to spend more than governments.

For America, nanotechnology is the largest federally funded science initiative since the country decided to put a man on the moon. In 2004, the American government spent $1.6 billion on it, well over twice as much as it did on the Human Genome Project at its peak. In 2005, it planned to shell out a further $982m. Japan is the next biggest spender, and other parts of Asia as well as Europe have also joined the funding race (see Chart 10.2). Perhaps surprisingly, the contenders include many developing countries, such as India, China, South Africa and Brazil.

In the six years up to 2003, nanotechnology investment reported by government organisations increased roughly sevenfold, according to figures from Mihail Roco, senior adviser for nanotechnology at America's National Science Foundation. This large

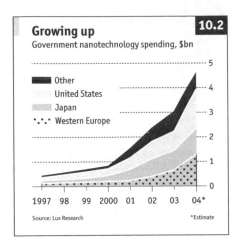

Growing up 10.2
Government nanotechnology spending, $bn

- Other
- United States
- Japan
- Western Europe

Source: Lux Research   *Estimate

amount of funding has raised expectations that may not be met. Some people worry that all the nanotechnology start-ups will help to inflate a bubble reminiscent of the internet one. But there are good reasons to think that the risk has been exaggerated. Private investors are being much more cautious than they were during the dotcom boom, and much of the money that is being spent by governments is going on basic science and on developing technologies that will not become available for years.

However, a number of existing products have already been improved through nanotechnology, with more to come in the next few years. Bandages for burns have been made antimicrobial by the addition of nanoparticles of silver. Fabrics have been stain- and odour-proofed by attaching molecules to cotton fibres that create a protective barrier. Tennis rackets have been strengthened by adding tiny particles that improve torsion and flex resistance. Other applications include coatings for the hulls of boats, sunscreen, car parts and refrigerators. In the longer term, nanotechnology may produce much bigger innovations, such as new kinds of computer memory, improved medical technology and better energy-production methods such as solar cells.

The technology's most ardent proponents claim that it will lead to clean energy, zero-waste manufacturing and cheap space travel, if not immortality. Its opponents fear that it will bring universal surveillance and harm the poor, the environment and human health – and may even destroy the whole planet through self-replicating "grey goo". Both sides overstate their case, but on balance nanotechnology should be welcomed.

Apply here

Where very small things can make a big difference

"**A**RGON! UP, UP! COME ON, BOY, LET'S GO!" Don Eigler, a researcher at IBM's Almaden Research Centre in California, is one of the world's experts in moving atoms. In 1989, he spelled out the letters "IBM" in xenon atoms, which made him the first person to move atoms individually. But today he is having trouble persuading his large Leonberger dog to get up from the office floor.

If dogs were as easy to move as atoms, Mr Eigler would be able to get Argon out of the office by using a computer mouse to point, click and drag him. But although the ability to move individual atoms is impressive, it is not particularly useful for anyone but experimental scientists like Mr Eigler. Most researchers think that moving atoms one by one will not be a practical way of creating new materials.

A better way may be to exploit the natural tendencies of atoms and molecules to crystallise, fold, form layers or otherwise self-assemble. Ordered molecular structures arise spontaneously, for example in crystallisation or in the formation of a snowflake. Scientists have already learnt how to use self-assembly to build nanoscale clusters of atoms, layers, pillars, tubes, ribbons, spheres, rods and rings, as well as more complex assemblies that resemble natural structures such as helices or even flowers. Much current research is concentrated on finding ways of arranging such nanoscale structures so they could serve as devices for, say, storing information or generating electricity from light to make solar cells.

Some, such as Harry Kroto, an eminent professor of chemistry at Britain's University of Sussex, say that nanotechnology is no more than advanced chemistry. But although so far much of nanotechnology has resembled traditional chemistry, increasingly the driving forces behind it are physics, engineering, materials science and information technology. These disciplines have brought new tools for working at the nanoscale, and for building in it too. They include the modern descendants of the scanning tunnelling microscope, and tools for writing, printing and even picking things up. Existing tools have also become more useful, having been combined with powerful computing, automation and visualisation methods such as virtual reality. All this work is allowing researchers to build libraries of new materials with different and useful properties, along with instructions on how to build them.

First catch your tube

In 1991, a researcher working at the NEC Corporation in Tsukuba, Japan, discovered a new form of carbon that turned out to have extraordinary properties. The so-called carbon nanotube is like a tiny sheet of graphite rolled into a cylinder, with a diameter of around a nanometre, and is very strong and light. It has become the star of nanotechnology. A host of uses has been proposed for it, including in sensors, molecular probes, computer memory, televisions, batteries and fuel cells. The list lengthens every time a new property is discovered in a slightly differently shaped or sized tube. In 2003 scientists at the University of Texas at Dallas found a way of spinning nanotubes into fibres to make the world's toughest polymer.

Meyya Meyyappan, head of nanotechnology at NASA, America's space agency, says that over the next few years an entirely new generation of flat-panel displays based on carbon nanotubes is likely to be developed. Field emission displays (FEDs) are based on the idea that electrons are emitted in a grid, rather than from a single source as in a television. Because electrons flow easily down the inside of a carbon nanotube, an array of these tubes could be used to draw the pixels on a display. It would be flat and more efficient than existing displays. NEC, a company that champions carbon nanotubes, is using carbon nanohorns (a close relation) as electrodes in a fuel-cell-powered laptop computer due for release in 2005.

Although carbon nanotubes are becoming cheaper, and can be manufactured in large quantities, it is still difficult to control their quality and purity, which for the moment limits the uses to which they can be put. But Richard Smalley, a professor of chemistry at Rice University in Houston, is convinced that within the next decade these problems can be overcome and production costs will drop.

However, some companies and researchers now hope that they will not have to wait that long, because it has emerged that a wide variety of other materials will also form nanotubes with useful properties. These may not be as potentially versatile as carbon ones, but for some applications – such as sensors, electrical components and lubricants – nanotubes or even nanowires made from other materials may be even better.

If just one nanoscale structure can have so many possible applications, it is clear why nanotechnology in general is causing such excitement. There are a host of other structures, and ways of putting them together, that also offer immense possibilities for new materials and devices.

One nanoparticle that is already being put to work is the quantum dot. Quantum dots are crystals made up of only a few hundred atoms. They can be produced from many materials and have the useful property that they can be made to fluoresce in almost any colour. Because they are so tiny – about the same size as a protein molecule or a short sequence of DNA – they can be used as probes to track reactions in living cells.

These tiny light tags are helping researchers in drug discovery, medical diagnostics and in the analysis of gene expression. As they can reveal exceptionally small quantities of biological molecules, they could be used in sensors that are better than existing technology based on DNA probes. So within a few years, quantum dots may begin to appear in over-the-counter diagnostic products.

Quantum Dot Corporation, of Hayward in California, launched its first product – a quantum dot attached to a specific biological molecule – in 2003. Since then, it has generated several million dollars of revenue, some from pharmaceutical giants such as AstraZeneca, Pfizer and GlaxoSmithKline.

There are plenty of other nanoscale technologies competing to build tiny sensors, for example to detect infection, which is likely to become much easier in the next few years. One technology uses minute particles of gold attached to DNA fragments that bind to the genetic material of pathogens, such as viruses or bacteria. When a sample of blood containing the pathogen is placed between two tiny electrodes, the gold particles close the circuit between the electrodes, revealing the presence of the pathogen.

John Ryan, a professor of physics at Britain's Oxford University, explains that nanotechnology can probe biological processes at the single-molecule level. This will be useful in all sorts of things from medicine to security, identifying tiny quantities of explosives, biological agents or even chemical weapons.

Another application for nanoparticles could be the enhancement of medical imaging. For example, iron particles might improve the quality of MRI scans. Nanoparticles could also be used to deliver drugs and genes to patients, allowing medicines to be taken in a more convenient form. In fact, there is a long list of areas in medicine that could benefit from the technology. After all, the constituents of human cells are also nanosized objects, so it seems likely that the tools and products of nano-technology will prove helpful.

Yet another important area likely to benefit from nanotechnology is information technology. Within a few years, nanotechnology could bring big improvements in the amount and types of computer memory available. And the new technologies being developed are likely, in the longer term, to bring big changes in computer logic chips.

Moore and more

People have got used to the idea that with each generation of computer chips, both performance and cost improve. This is done by shrinking components and cramming in more transistors per square inch. But for some time now the pundits have been saying that Moore's law, as the process is known, could not go on for ever.

A decade ago they predicted that the complementary metal-oxide semiconductor (CMOS) process, by which the vast majority of the world's silicon chips are made, would reach its limits at 100nm. Today, all the big chip manufacturers, including Intel and Transmeta of Santa Clara, California, and AMD of Sunnyvale, are producing parts with nanoscale features. Paolo Gargini, Intel's director of technology strategy, said that by the end of 2005 he hoped to be producing parts with 35nm components, and his laboratories can make features as small as 10nm.

Whether or not chipmakers are already in the realm of nanotechnology is a matter of debate. True nanotechnology must take advantage of the novel properties that arise at that scale; simply making things smaller does not count. But Intel has demonstrated a new technique – for pulling apart silicon atoms to speed up the flow of electrons – that would surely qualify.

Chip manufacturers have been ingenious at prolonging the era of the silicon chip, and Dr Gargini thinks they can continue doing so for about another decade. It is not so much the difficulty of making things increas-ingly small but the cost that is the limiting factor, because every genera-tion of chip fabrication requires a bigger injection of capital to build more precise manufacturing tools. So if self-assembly can be used to put together logic and memory chips from the ground up, nanotechnology will come into its own.

Companies such as Intel have a multibillion-dollar capital investment in their production equipment. They would rather use nanotechnology to extend the life of their chips than build entirely new ones. One possi-ble technique might be to grow tiny tubes and wires made from silicon, germanium or carbon on the surface of chips to allow electrical charge to flow with less heat. This, says Dr Gargini, might extend the life of CMOS to 2015–25.

Logic chips are much more difficult to make than those that store memory. They have to do complicated things such as adding, subtracting and multiplying, whereas memory merely has to store information in neat rows. Because making memory chips is simpler, and because lots of different companies are working on ways of harnessing nanotechnology for this purpose, a revolution in this area is closer at hand. Hewlett-Packard (HP), a Silicon Valley veteran, is only one of the companies looking at creating a new type of memory, using individual molecules as components in switches and transistors.

Philip Kuekes, a researcher at the firm, thinks it will be cheaper to design and build devices taking advantage of effects that become predominant at the quantum scale than to try to overcome those effects with devices designed at larger scales. He is working on chemical tricks to lay down a regular grid of nanoscale wires only a few atoms thick. Between two layers of these grids will be one layer of switches a single molecule thick. If a logic chip could be made this way, it would hold tens of billions of logic elements, compared with 50m on existing chips.

But that is only one of a range of promising technologies. In Zurich, IBM is building a memory device known as Millipede that can store data at such density that 25m printed textbook pages could be put on to a postage stamp. Such developments promise non-volatile memory (the kind that retains information with the power off) so large that computers may no longer need hard drives.

Lighting-up time

The last big area tipped to benefit from nanotechnology is energy, both through its more efficient use (particularly in lighting) and through more effective ways of generating electricity. Clayton Teague, director of the National Nanotechnology Co-ordination Office in Arlington, says that nanoscale particles used in new solid-state lighting could cut the electricity used for illumination by up to 50% by 2025. Ordinary light bulbs would be replaced with improved versions of light-emitting diodes (LEDs) that emit bright white light.

At General Electric, researchers are trying to improve the structure of phosphorus to make fluorescent lighting more energy-efficient. And Cerulean International in Oxford is marketing a nanoparticulate diesel additive that gives an improvement in fuel economy of up to 10%. Many other companies are working on better catalysts via nanostructured materials.

Nanotechnology should also bring energy savings from more streamlined manufacturing. Tyler McQuade, a researcher at Cornell University, is working on ways of simplifying complex multi-stage drug manufacturing processes by encapsulating the different chemicals involved in nanoscale spheres. He thinks he can manufacture Prozac in a single step with little waste. Pfizer, he says, generates 25kg of waste for every 1kg of product.

Nanotechnology may also help to bring energy technologies such as fuel cells to market. Cheap and efficient solar cells look within reach, using newly developed materials to replace the fragile and expensive silicon-based wafers currently in use. Researchers in companies such as Nanosolar, in Palo Alto, are developing materials that convert light into electricity and can be sprayed or printed on to a flexible plastic sheet.

Nanotechnology is still in its infancy, although over the next few years an increasing number of products are likely to incorporate it in some way. But some of the longer-term benefits of the discoveries made today will not become apparent for decades. Mr Meyyappan at NASA tells the story of Herbert Kroemer, who 40 years ago was working on an obscure bit of semiconductor theory at America's RCA. No one, not even Mr Kroemer himself, had the slightest idea that one day this work would lead to a technology that has become ubiquitous in lasers in everyday objects such as CDs and DVDs.

Fear and loathing

Some of the worries about nanotechnology are rational, some not

NOT LONG AGO ELLA STANDAGE was woken by a bad dream about nanobots. She was terrified that nanoscale machines might replicate uncontrollably and turn the entire planet into grey goo. Ms Standage is not the only one to worry about such imaginary horrors, but at least she has an excuse: she is only four years old.

The grey-goo idea goes back to a prediction by Eric Drexler, chairman of the Foresight Institute, a nanotechnology-policy group in Palo Alto, that one day all manufacturing would be done by very tiny robots. He thought that given the correct plans, these minute devices would be able to produce any item – an engine, say, or a pair of jeans – from nothing more than a feedstock of atoms.

For this plan to work, though, these robots would have to be able to make more of their own kind, otherwise things would take far too long to build. Mr Drexler thought these hypothetical nanobots would have to be self-replicating, and gave warning that care would have to be taken to ensure they did not replicate out of control. This idea launched a wave of public concern. If these nanobots started making copies of themselves by scavenging materials from their environment, it was suggested, they would eventually become visible to humans as a seething mass of tiny robots, or grey goo – and might ultimately consume the entire planet.

Little wonder that nanobots have become a favourite new bogey in everything from children's cartoons (where Ms Standage learned of them) to films. In vain do scientists protest that as yet nobody knows how to build a self-replicating robot of any size. They have also put forward various theoretical reasons why there could never be such a thing. For example, they ask, how would those robots get the energy to survive? But lack of feasibility is no obstacle to the imagination, and the idea of nanobots fits well with contemporary fears about out-of-control science.

Frankenparticles

However, interest in grey goo seems to be on the wane as more serious concerns about the potential toxicity of nanoparticles are emerging. It is already clear that some manufactured nanoparticles are harmful to mice and fish. Ken Donaldson, a professor of respiratory toxicology at the University of Edinburgh, says that nanoparticles and nanotubes are likely to be far more toxic than the same chemical in larger form because smaller particles have a greater surface area and are far more reactive. Being so tiny, these particles may be able to penetrate cells and evade the human immune system. When inhaled, says Mr Donaldson, fine carbon particles are able to enter the blood and the brain. In 2004 Swiss Re, an insurance company, published an alarming report on the possible dangers of nanoparticles and the liability issues that could arise from products using them.

However, a report in the same year from Britain's Royal Society said that in most cases people's exposure would be limited: where nanoparticles had been used as raw materials, ingredients or additives in a product, they would usually be contained within a composite or fixed to a surface. Indeed, nanoscale materials have been used for years, for example in computer hard drives, without causing any problems. If governments, industry and scientists continue to take the issue seriously, creating new nanoparticles seems no more risky than creating new chemicals.

Besides, nanoparticles are already all around us: the air is full of them, from the exhaust of diesel engines, cigarette smoke, hairspray, burning candles and toast. People create and use all sorts of nasty toxic chemicals every day. Moreover, many willingly have their faces injected with botox, a highly toxic substance, purely to indulge their vanity.

In fact, toxic nanoparticles will have their uses. Scientists are already trying to wrap them in harmless coatings so they can be used to fight diseases or destroy cancerous cells. The tailored delivery of a toxic nanomaterial to treat cancer would be far preferable to existing methods that flood a patient's entire body with toxins. The ability of nanoparticles to cross the blood-brain barrier may also turn out to be useful because at present it is very difficult to get drugs across that barrier. Yet more information about the toxicology of these materials is urgently needed, and a great deal of work is already under way.

Even so, one small but vocal anti-nanotechnology group, the Action Group on Erosion, Technology and Concentration (ETC), has called for a complete moratorium on the use of synthetic nanoparticles such as quantum dots, nanowires and nanotubes. ETC, based in Canada, is a small fringe group of activists, but hogged the headlines when its concerns were picked up by Britain's Prince Charles.

Most advocates of nanotechnology say that a ban would stop research to assess and mitigate any risks. In June 2004, a group of nano-technology experts from 25 countries met in Virginia under the auspices of America's National Science Foundation to discuss how nanotechnology could be developed responsibly. Most of them agreed that no moratorium should be imposed because it would prevent any risks from being assessed.

Even Jim Thomas, ETC's programme manager, who is based in Oxford, thinks this aspect of nanotechnology is "a manageable issue", and hopes that the debate will not be about health alone. What really troubles ETC is "corporate concentration": that companies involved in nanotechnology are acquiring too much power. And ETC is not alone. In July 2004 a British pressure group, Corporate Watch, launched a project to map the nanotechnology industry in order to provide campaigners with "individual corporate targets". In some of its literature, under the headline "Nanotech is Godzilla", the group talks about the "dark side of nanotech: hazardous substances, military applications and a huge leap in corporate power". Nanotechnology, like GM, seems to have become a handy tool for corporation-bashing.

Do such fringe groups matter? The lesson of the debate over genetically modified (GM) food was that they can be crucial in influencing broader public opinion, especially where there are worries about human health and the environment. Some of the groups that were involved in successful protests against GM are spoiling for a fight over nanotechnology too.

Pay attention

Moreover, there are signs that some of these concerns are being taken up by more mainstream voices. James Wilsdon, head of strategy at Demos, a British-based think-tank, thinks that nanotechnology needs to be "opened up for discussion", and questions should be asked such as, "What is the technology for? Who controls it? Who will take responsibility if things go wrong?" And Britain's Cambridge University recently recruited a staff ethicist at its Nanoscience Centre. The head of the unit, Mark Welland, says this is an experiment, aimed partly at ensuring that their scientists take ethical concerns on board. But it also allows the unit to engage with groups such Demos and Greenpeace, and cleverly allows it to be part of the debate rather than its subject.

Many see parallels between GM and nanotechnology, and there have been warnings that the public could reject nanotechnology, as it did GM in Europe. But there are good reasons to think that this time the response will be different.

The main one is that on nanotechnology the views of pro-environment, pro-poor and anti-corporate groups are not aligned. For example, two big environmental groups, Environmental Defence and Greenpeace, are cautiously optimistic about the technology. Doug Parr, chief scientist at Greenpeace, has some concerns, including the health and environmental risks of nanoparticles and its potential use for military purposes. But the very scope of nanotechnology makes it difficult to oppose. As Mr Parr says, "We still don't have a policy on nanotechnology; you cannot on something that is so diverse." And he adds, "Increasingly we recognise some good things can come from it." One of the things Mr Parr has in mind is that new materials might bring down the cost of solar cells and thus make solar energy viable and even lucrative.

In fact, nanotechnology has many potential environmental benefits, which makes it hard for green groups to oppose it in principle, as they did with GM. Besides those energy savings, specialised nanoparticles or porous materials might be used to detoxify polluted water, land and even air. And greens can hardly accuse the technology of trying to do something "unnatural" when humans have been modifying substances to create new materials since the Bronze Age.

Pro-poor groups already worry that new materials might result in big changes in demand for commodities such as copper, cotton or rubber, but they too will find nanotechnology hard to oppose because of the benefits it may bring. For example, vaccines might be encapsulated in nanomaterials so that they no longer need to be refrigerated, and water desalination could be made cheaper.

Groups concerned with developing countries are also worried about maintaining access to such technology, pointing to the race now in progress to buy up rights to the key areas of nanotechnology in the hope of bagging a valuable future patent. This may well be a problem, although not one that is unique to nanotechnology. But the battle is not being fought along traditional rich-versus-poor lines; indeed many developing countries are heavily engaged in nanotechnology. Among the more unexpected countries on the list are China, India, South Korea, Brazil, Chile, Argentina, the Czech Republic, Mexico, Romania, Russia and South Africa. Even little Costa Rica is investing in this area.

Because the technology is so new, all these countries see an opportunity for getting a slice of the action, as well as a way of solving long-standing problems. One Indian group is working on a prototype kit for detecting tuberculosis, and the Chinese have developed a scaffold for broken bones that is now being tested in hospitals.

For pro-poor groups, the fight against GM was essentially a fight against America, and more specifically against Monsanto, the company that came to symbolise GM. But nanotechnology already involves a wide range of participants, including the governments of many poor countries. A pro-poor group would probably not be able to exercise much leverage against the nanotechnology research of a country such as China.

Fears that the public might reject nanotechnology have allowed some groups to try a new tack: telling scientists, companies and governments that if they want their technology to be widely accepted, they must "democratise" it. What exactly that means is not clear, except that they are trying to harness public opinion to serve their own particular agenda.

In fact, nobody really knows what the public wants from nanotechnology. According to two surveys in America and Britain, most people do not even know what it is. And although they are unlikely to reject it outright once they find out, pressure groups will certainly be able to sway public opinion on some aspects of it. Companies working on nanotechnology applications in new products will need to bear that in mind.

Downsizing

Companies both large and small hope to make big money from tiny particles

A T THE PALO ALTO HEADQUARTERS of Nanosys, a nanotechnology start-up, Stephen Empedocles, the director of business development, is demonstrating some of his gee-whizz technology. His exhibit is a flat disc, where the surface on one side has been designed with a structure at the nanoscale that repels water. Dr Empedocles puts a few drops of water from a pipette on to the non-modified side. The droplets bead and cling as they would on the surface of a polished car. On the hydrophobic side, however, the droplets bounce and roll off like high-speed ball-bearings.

Nanosys is one of the most talked-about start-ups in nanotechnology, but not for the reasons it would like. What it wants people to discuss is its library of materials with pre-determined characteristics (such as hydrophobic surfaces), made from an array of proprietary nanostructures. Instead, it has been getting lots of publicity for its attempt to go public in 2004. Its offering of 29% of its shares, at $15–17 a share, would have valued the company at well over $300m, but the issue was pulled because of poor market conditions at the time.

One of Nanosys's main assets is its intellectual-property portfolio, which is broader than that of most nanotechnology start-up companies. It was this portfolio that the market was being asked to value in the summer of 2004. The company does not think that potential investors were put off by its filings with America's Securities and Exchange Commission, which said that it did not expect any products to emerge for several years, if ever, and they "may never achieve profitability".

However, some in Silicon Valley think it was a good thing that the initial public offering did not succeed. Scott Mize at the Foresight Institute, a pro-nanotechnology think-tank, says the company tried to go public too early and its lack of products and revenues would not have supported stock prices.

Vinod Khosla, a partner in Kleiner Perkins Caufield & Byers (KPCB), a large venture-capital house, says the most important requirement for any technology IPO is predictability. Revenue is less important, he says, as long as the firm can convince investors that it will arrive in time. By

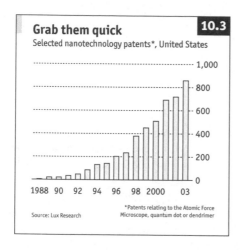

Grab them quick `10.3`

Selected nanotechnology patents*, United States

1988 90 92 94 96 98 2000 03

*Patents relating to the Atomic Force Microscope, quantum dot or dendrimer

Source: Lux Research

and large, argues Mr Khosla, there is not enough accumulated experience in nanotechnology to be able to tell how long it will take to get products to market.

One reason for that uncertainty is a huge proliferation of patents. Between 1976 and 2002, about 90,000 nanotechnology patents were registered with America's patent office alone. Some analysts have issued warnings about an "intellectual-property land-grab" and predicted trouble ahead because of the breadth, and overlapping nature, of some of the patents.

Matthew Nordan, at Lux Research, a nanotechnology consultancy in New York, sees an impending war over patent infringements where lawsuits will be flying. The areas most likely to be affected are carbon nanotubes and quantum dots. IBM, for example, holds a key bit of intellectual property on a method of producing carbon nanotubes. The company is rumoured to have more lawyers than engineers working on nanotechnology. But for the moment there is no point in anyone suing, because nobody has yet made any real money from these nanoparticles.

Dynamite in small packages

If Nanosys's IPO had been successful, it might have paved the way for a number of other companies waiting in the wings. There is much excitement about nanotechnology's potential for existing and new businesses, but nobody wants to see this potential ruined by early hype.

Some also feel jittery about another technology-driven stockmarket bubble so soon after the dotcom bubble. The most basic things driving speculation – fear and greed – have not gone away. And there seem to be plenty of private investors who view the dotcom era as a missed opportunity and want something, anything, to invest in. So far, though, despite the billions of dollars being spent on nanotechnology research, there are only a handful of public companies in the sector, and most of these are small.

Josh Wolfe, co-founder of Lux Capital, a venture-capital firm, and editor of the *Forbes/Wolfe Nanotech Report*, says a number of firms are "nanopretenders" – companies that put a "nano" in their names but actually do things at larger scales. One of them is Nanometrics, a company in Milpitas, California, that makes tools at the microscale (1,000 times larger than the nanoscale). Another is Nanogen, in San Diego, which creates gene chips far larger than the nanoscale. Its share price has been volatile.

Even professional investors can get the sector wrong. Merrill Lynch, an investment bank, was left red-faced after the launch of its nanotechnology tracking index in April 2004 when some of the firms it had picked turned out not to be nanotechnology companies after all and had to be dropped. At the time the index was announced, every company in it got a boost, but six months later share prices were down by a quarter. The same thing happened to the shares in a nanotechnology index launched by Punk Ziegel, another investment bank.

At KPCB, Mr Khosla is worried about indices. "When companies like Merrill Lynch start having a nanotechnology index, I think that is getting into the hype cycle for which a lot of people got into a lot of trouble during the internet bubble." When people start getting interested, he says, fund managers decide that some proportion of their investment should be in nanotechnology. A bubble gets going when everybody starts piling in, trying to buy stocks that do not exist, he adds. "To me, an index is just an example of hyping."

If there is one thing everyone agrees on, it is that nanotechnology is neither an industry nor a market. Lumping together different nanotechnology firms may be as sensible as assembling a group of firms whose names start with Z. A company selling nano-improved fabrics has little in common with one developing solar cells. To add to the problems, some of these indices include big companies for which nanotechnology is only one of many activities.

It is easy to see why a nanotechnology bubble might form, but if so, it will be nothing like as big as the ill-fated internet one, for several reasons. One of them is offered by Steve Jurvetson at Draper Fisher Jurvetson, a venture-capital firm based in Menlo Park, California: the number of people who can enter the business is limited by the number of science graduates available. In America, there is currently a shortage of science PhDs. Business school graduates working in banking or consulting cannot start nanotechnology companies in the way they created new internet companies.

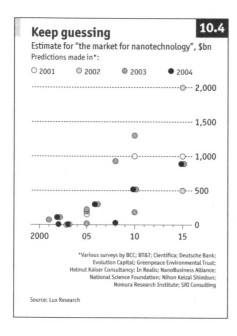

Keep guessing 10.4

Estimate for "the market for nanotechnology", $bn

Predictions made in*:

○ 2001 ◔ 2002 ◉ 2003 ● 2004

*Various surveys by BCC; BT&T; Científica; Deutsche Bank; Evolution Capital; Greenpeace Environmental Trust; Helmut Kaiser Consultancy; In Realis; NanoBusiness Alliance; National Science Foundation; Nihon Keizai Shimbun; Nomura Research Institute; SRI Consulting

Source: Lux Research

Another anti-bubble factor is the high capital cost of setting up business in nanotechnology. Start-ups need many millions of dollars to pay for equipment, proof of concept and salaries for highly trained staff. Mr Jurvetson's investment firm specialises in investing in early-stage nanotechnology companies. All of them, he says, are spun out of university or government labs, not Silicon Valley garages.

And venture capitalists appear to be in no mood to fund flaky proposals. Indeed, venture-capital funding in nanotechnology actually declined between 2002 and 2003. At one nanotechnology get-together in 2004, venture capitalists complained that there was more investment capital available than there were good ideas to fund.

Take your pick

One frequently repeated nugget of wisdom in the nanotechnology business is that the real money is in making the picks and shovels, as it was in an earlier Californian gold rush. The pick-and-shovel manufacturers in nanotechnology are making tools such as microscopes, manipulators and instruments for working at the nanoscale. They include microscope-makers such as Veeco Instruments of Woodbury, New York; FEI of Hillsboro, Oregon; and smaller companies such as Infinitesima of Bristol, England.

Everyone working in nanotechnology, whether at the lab bench or at a chip-manufacturing company, needs to see where they are putting their atoms. But there are also companies developing new lithography processes, for writing and printing at the nanoscale, which are likely to be important in manufacturing.

So how big is the nanotechnology industry as a whole? There are plenty of estimates, but they vary from tiny to huge (see Chart 10.4).

Nano a-go-go	10.5

Some interesting nanotechnology companies to watch

Company	Technology
Konarka, Lowell, MA	Photovoltaics
Nantero, Woburn, MA	Non-volatile RAM using carbon nanotubes
QinetiQ Nanomaterial, Farnborough, UK	Specialty chemicals and nanopowders
Dendritic NanoTechnologies, Mount Pleasant, MI	Nanoparticles for drug encapsulation, delivery and release
Quantum Dot, Hayward, CA	Nanocrystals for life-science research
Frontier Carbon Corporation, Tokyo, Japan	Specialty chemicals and carbon nanotubes
Molecular Imprints, Austin, TX	Nanolithography
NanoInk, Chicago, Il	Dip pen lithography
Zettacore, Denver, CO	Molecular memory
Zyvex, Richardson, TX	Tools, materials and structures for research and manufacturing

Source: Company websites

Mr Nordan at Lux Research explains why: "The conventional wisdom here is that there's something called the nanotechnology market that has things in it called nanotechnology companies that sell nanotech products. All of these things are wrong."

That has not stopped him from trying to quantify this non-existent entity in a report. He divides it into three parts: nanomaterials (tubes, particles, dots), nanointermediates (products made of these materials such as films, memory, fuel cells, solar cells) and nano-enabled products (such as the Chevrolet Impala car with body mouldings made of nanocomposites). In 2004 these three sectors added up to $158 billion in product revenue, most of it from the semiconductor industry.

In the next decade, says Mr Nordan, nanotechnology will be incorporated into products worth $2.9 trillion, and most of this revenue will be from new and emerging nanotechnology. But such estimates have to be treated with caution. They would include a $30,000 car with $200 side panels improved by nanotechnology.

All the same, this kind of work is useful because it gives an indication of where money could be made. Most nanomaterials, says Mr Nordan, will rapidly become commodities, with operating margins in the high single digits – the sort of figures typical of specialty chemicals. Ten years from now he would expect this business to be worth about $13 billion – a tiny share of the market for materials in general. He says that new materials will not be able to command large margins because that would make the economics for downstream manufacturers unattractive. Margins on nanointermediates and nano-enabled products, he predicts, will be similar to those in traditional product categories. For example, the margins on drugs will be far higher than those on clothing.

Which particular companies are likely to do well? Traditional wisdom has it that start-up companies are the most likely to discover the new technologies that will offer better performance and lower costs. Big companies such as Intel and Sharp, with multibillion-dollar investments in plant and equipment, are thought to be resistant to revolutionary change, whereas small companies can win by using entirely new, "disruptive" technologies that make large existing investments redundant.

What is different about nanotechnology is that the next disruptive idea could quite easily come from a big company rather than a small one. Large firms are well aware that the technology will provide the basis for many new and improved products in the future, and are investing heavily to stake their claims. But they often choose to keep quiet about it.

Sometimes their nanotechnology research programmes take them into unfamiliar territory. General Electric, for example, is looking at iron nanoparticles that might be useful in medical imaging. Future discoveries could come from unexpected places, because a patent on one aspect of a nanoscale structure could be useful in applications from IT to pharmaceuticals.

In the next decade and beyond, nanotechnology will bring wave upon wave of new discoveries. Some IPOs have been delayed by the Nanosys hiccup, but there were not that many waiting in the wings. Many of the interesting new companies are still several years away from going public. And some of these may well be bought up by bigger companies, rather than coming to the market.

Some people think that nanotechnology is likely to lead to massive social, economic and technological changes. If its effects could be that important, should governments control and regulate it?

Handle with care

Nanotechnology promises great benefits, but safeguards will be essential

IN THE EARLY 1800s, groups of English workers wrecked machines that they felt threatened their jobs. They were called "Luddites" after one of their leaders, a term that is now used for anyone who puts up resistance to new technologies. The odd thing about nanotechnology's Luddites is that they have started resisting before the technology has really established itself.

As people start to buy products involving nanotechnology, from odour-resistant shirts to window glass that repels dirt, they will realise that many of these new things are useful and harmless. And as awareness of nanotechnology grows, they will begin to understand that it covers a range of different ways of doing things, of which some carry some risk and others do not. As a result, the technology's detractors will probably become more nuanced in their complaints.

Nanotechnology has the potential to cause an industrial upheaval, just as electricity did in its time. Like electricity, though, it has so many and such diverse applications that it is unlikely to arrive in one huge wave, as nanotechnology's critics fear. Instead, there will be a series of smaller waves. Many of the innovations the technology may bring are a long way off, leaving plenty of time to prepare.

Fuzzy round the edges

That is just as well, because in the longer term some awkward questions will have to be answered, mainly in an area known by the clumsy name of "convergence between nanotechnology, biotechnology, information technology and areas of cognitive science" (NBIC). Scientists have noticed that the divisions between these areas are disappearing. For example, if a new nanoparticle is inserted into a cell, is this biotechnology or nanotechnology? If this molecule has a memory that can record events taking place in the cell, is this nanotechnology or computing?

If artificial molecules can be designed to serve as memory and logic, it might be possible to put a computer inside a cell. This could monitor and modify the way it works – for example, by detecting a molecule that might cause a disease, and taking action to ward it off. Scientists have already put together a primitive form of this kind of DNA computer in a test tube.

That also means, though, that one day it might be possible to program in enhancements to humans at cellular level. This is likely to cause concern. People will probably also worry about the emerging possibility of forming direct connections between machines and the human brain. And if convergence in these technologies could bring about a big increase in the human lifespan, that could have a profound effect on the nature of society.

For the moment, none of these futuristic things can be done. The more mundane applications available so far do not require new regulations in countries that already have strong legislation in areas such as chemicals, health and safety in the workplace, pharmaceuticals and the environment. However, it would be sensible for governments to examine all existing regulations to make sure that they provide adequate cover for the new products of nanotechnology.

For example, most countries require a new chemical to be assessed before it can be sold. But nanoparticles of an existing chemical may have properties so different from the bulk form that new nanoparticles should be treated as if they were new chemicals. Regulations may assume that the nanoscale version of a chemical behaves in the same way as it does in bulk, which it may not. A carbon nanotube could be thought of as graphite, but it also has a structure that resembles asbestos. It is essential to find out whether these tubes behave like pencil lead or like a highly dangerous mineral.

Other regulations may assume that the toxicity of a substance is directly related to the quantity that is present. Again, this may not be true. Nor is it safe to rely on rules that say companies need report only the known risks of a substance. Many nanoparticles are completely new to science. They are being developed at a very rapid rate, so they may present risks that nobody is even looking for yet.

Ambiguities in existing regulations will also have to be resolved. For example, medical devices and drugs may be covered by separate and quite different legislation. But is a nanoparticle that fights cancer a drug or a device? Even if, technically, it is classified as a device, most people would expect a cancer-fighting nanoparticle that they swallow to be regulated as a drug. Consumers would like to think that governments are reviewing existing legislation to make sure that special issues raised by nanotechnology are properly covered.

Companies, for their part, need to be open about the nanoparticles they are putting in their products, and about the kind of testing that has been done. There may be no reason to worry about such ingredients, but secrecy over testing is already raising suspicions among the technology's critics.

Patents are another area that needs careful thought. If most of the huge volume of new patents is coming from universities, it is reasonable to ask how this intellectual property is licensed, and whether researchers and potential users in developing countries might be granted greater access, perhaps by making it free for research purposes and cheap for poor countries.

There is also the question of privacy. New cheap and small sensors may make it much easier to monitor things such as health and environmental conditions. For example, it may be possible to detect cancers when they are still tiny. That is all to the good. But a future generation of cheap, highly effective and ubiquitous sensors may also make it possible, say, to screen large numbers of people to find out if they have consumed too much alcohol. Such applications might be more controversial.

You can't vote on everything

The idea of "democratising" nanotechnology – giving ordinary people more of a say in what areas of science and technology should be pursued – is unlikely to be helpful. For a start, it is next to impossible to slow down or control some areas of science in one country when the world is so interconnected. Just look at the attempts to reach an international agreement to ban human cloning: many countries do not want a ban if it also prevents research involving therapeutic cloning which they do want, and different countries may take different views on whether something is useful or ethical.

Nanotechnology, like any new discovery, offers both risks and rewards. There will undoubtedly be some need to control its exploitation to minimise the risks, but there are also strong arguments for allowing the unfettered pursuit of knowledge: without it, innovation cannot flourish.

Twenty years ago, nobody could have foreseen that the invention of a new microscope would launch a remarkable new technology, perhaps a revolution. Scientists should be allowed to work with as little hindrance as possible to gain a better understanding of the object of their study – however large or small.

The material on pages 306–29 first appeared in a survey in *The Economist* in January 2005.

11
ROBOTS AND ARTIFICIAL INTELLIGENCE

The gentle rise of the machines

The science-fiction dream that robots would one day become a part of everyday life was absurd. Or was it?

W HO WOULD HAVE THOUGHT that a Frisbee-shaped contraption that extracts dust from carpets would be the state of the art in household robots at the dawn of the 21st century? Since its launch in 2002, Roomba, a circular automatic vacuum cleaner made by a firm called iRobot, has swept up millions of dollars from over 1m buyers. Rival firms such as Electrolux and Karcher sell similar but pricier sweepers. Robot vacuum cleaners, it seems, are catching on.

Are these mere playthings, or the beginning of a new trend? Roomba is just the tip of the iceberg, according to Helen Greiner, co-founder of iRobot, which also sells industrial and military robots. Dan Kara of Robotics Trends, a consultancy, agrees. "The tipping point might be Roomba," he says.

Even if this is true, however, it would be quite a come-down compared with the robotic future that seemed, for much of the 20th century, to be just around the corner. Since 1939, when Westinghouse Electric introduced Electro, a mechanical man, at the World's Fair in New York, robot fans have imagined a world filled with tireless robotic helpers, always on hand to wash dishes, do the laundry and handle the drudgery of everyday tasks.

So far, however, such robots have proliferated in science fiction, but have proved rather more elusive in the real world. But optimists are now arguing that the success of the Roomba and of toys such as AIBO, Sony's robot dog, combined with the plunging cost of computer power, could mean that the long-awaited mass market for robots is finally within reach. "Household robots are starting to take off," declared a recent report from the United Nations Economic Commission for Europe (UNECE). Are they really?

Hand-built by robots

Although the dream of the home robot has not died, robots have had their greatest impact in factories. Industrial robots go back over 40 years, when they first began to be used by carmakers. Unimate, the first industrial robot, went to work for General Motors in 1961. Even at a time when computing power was costly, robots made excellent workers and proved that machines controlled by computers could perform some tasks better than humans. In addition, robots can work around the clock and never go on strike.

There are now about 800,000 industrial robots around the world, and orders for new robots in the first half of 2003 were up a record 26% from the same period in 2002, according to the UNECE. Demand is increasing as prices fall: a robot sold in 2002 cost less than one-fifth of an equivalent robot sold in 1990, for example. Today, in car factories in Japan, Germany and Italy, there is more than one robot for every ten production workers.

Similarly, agricultural robots harvest billions of tonnes of crops every year. There are six-legged timber cutters, tree-climbing fruit-pickers, robots that milk cows, and others that wash windows, trucks and aircraft. Industrial robotics is a $5.6 billion industry, growing by around 7% a year. But the UNECE report predicts that the biggest growth over the next few years will be in domestic rather than industrial robots. Sales of such devices, it predicts – from toys to lawnmowers to, yes, vacuum cleaners – will grow ten-fold between 2002 and 2006, overtaking the market for industrial robots.

The broader application of robotics is becoming possible thanks to the tumbling cost of computing power, says Takeo Kanade of Carnegie Mellon University's Robotics Institute, who has built robots on both sides of the Pacific. This lets programmers write more sophisticated software that delivers more intelligent robotic behaviour. At the same time, he notes, the cost of camera and sensor chips has tumbled too. "The processing power is so much better than before that some of the seemingly simple things we humans do, like recognising faces, can begin to be done," says Dr Kanade.

While prices drop and hardware improves, research into robotic vision, control systems and communications have jumped ahead as well. America's military and its space agency, NASA, have poured billions into robotic research and related fields such as computer vision. The *Spirit* and *Opportunity* rovers exploring Mars can pick their way across the surface to reach a specific destination. Their human masters do not specify the route; instead, the robots are programmed to identify and avoid obstacles themselves.

"Robots in the first generation helped to generate economies of scale," says Navi Radjou, an analyst at Forrester, a consultancy. Now, he says, a second generation of more flexible and intelligent robots will be able to do many more jobs. Hence the UNECE report's suggestion that domestic service robots might now be entering "into a diffusion process similar to that which the PC, the mobile telephone or the internet have had in recent years". But if robots really are poised on the cusp of ubiquity, what will they be used for?

Robots go home?

A possible robotic foot in the door could be toys. For the past few years, robots have been among the bestselling toys in the world. And they can be more than just playthings, once they have been hooked up to a network. Personal robots, wireless systems and cheap cameras, all tied together by a PC, could enable robots to water the plants while you are on holiday, or provide a roving set of eyes and ears. Sony's robot dog, AIBO, for example, can be linked wirelessly to a PC, so you can remotely monitor your home through its eyes as it walks around.

Another possibility, long touted by robot fans, is the use of robots to provide nursing care and assistance to the old and infirm. Honda, Mitsubishi and scientists at the Korean Institute of Science and Technology are designing machines to help old or disabled people move from room to room, fetch snacks or drinks, operate the television, and even call the doctor when needed. Though it has been notoriously difficult to pull off, Joe Engelberger, the inventor of Unimate, feels care of the elderly is precisely the opportunity the robotics industry should be pursuing. "Every highly industrialised nation has a paucity of help for vast, fast-growing ageing populations," he says.

Given that homes are designed for human inhabitants, the best shape for such robots might be humanoid. In Japan, the development of such robots – by firms such as Honda, Mitsubishi and Toyota – seems to have become a symbol of technological superiority. But ultimately, says Mr Engelberger, who went through this with Unimate, if domestic robots are going to succeed, they will have to be reliable and demonstrate value for money. You have got to be able to show, he says, "how this damn thing can justify itself".

They are already among us

Yet for all the progress in computing, there has not been a corresponding leap forward in robotics. Talk of robot helpers for the elderly has been around for years. Only a fervent optimist would take the success of the Roomba as the dawning of a new robotic era. But there is another way to look at things. We may, in fact, be surrounded by more robots than we realise. The trouble is that they have not taken on the forms that Hollywood, or robot researchers, led us to expect. Automated machines have, however, quietly slipped into many corners of everyday life.

Far more prevalent than robot vacuum cleaners are copiers that collate, staple and stack your documents and automated-teller machines that, as their name suggests, save human bank tellers the trouble of dispensing cash. Other machines scan groceries, wash dishes, make bread, sort mail by reading hand-written addresses and dispense train tickets. Commercial airliners fly and even land themselves using radar and satellite-positioning systems to navigate through fog and storms. Autonomous trains, akin to giant robotic snakes, drive themselves. All of these devices are autonomous computer-controlled machines, capable of responding to changing circumstances in accordance with orders from their human masters. They are, in other words, robots. But they are not the general-purpose mechanical men that most people associate with the term.

Why not? The answer, ironically, could lie in the rapid advance of computing power. Back in the mid-20th century, when the robotic future was being imagined, computers were huge and expensive. The idea that they would become cheap enough to be integrated into almost any specialised device, from a coffee-maker to a dishwasher, was hard to imagine. Instead, it seemed more likely that such intelligence would be built into a small number of machines capable of turning their robotic hands to a range of different tasks. In place of the general-purpose housebot, however, we are surrounded by dozens of tiny robots that do specific things very well. There is no need to wait for the rise of the robots. The machines, it seems, are already among us.

The material on pages 332–5 first appeared in The Economist in March 2004.

AI by another name

After years in the wilderness, the term "artificial intelligence" seems poised to make a comeback

LIKE BIG HAIRDOS and dubious pop stars, the term "artificial intelligence" (AI) was big in the 1980s, vanished in the 1990s – and now seems to be attempting a comeback. The term re-entered public consciousness most dramatically with the release in 2001 of A.I., a movie about a robot boy. But the term is also being rehabilitated within the computer industry. Researchers, executives and marketing people are using the expression without irony or inverted commas.

And it is not always hype. The term is being applied, with some justification, to products that depend on technology that was originally cooked up by AI researchers. Admittedly, the comeback has a long way to go, and some firms still prefer to avoid the phrase. But the fact that others are starting to use it again suggests that AI is no longer simply regarded as an overambitious and underachieving field of research.

That field was launched, and the term "artificial intelligence" coined, at a conference in 1956 by a group of researchers that included Marvin Minsky, John McCarthy, Herbert Simon and Alan Newell, all of whom went on to become leading lights in the subject. The term provided a sexy-sounding but informative semantic umbrella for a research programme that encompassed such previously disparate fields as operations research, cybernetics, logic and computer science. The common strand was an attempt to capture or mimic human abilities using machines. That said, different groups of researchers attacked different problems, from speech recognition to chess playing, in different ways; AI unified the field in name only. But it was a term that captured the public's imagination.

Most researchers agree that the high-water mark for AI occurred around 1985. A public reared on science-fiction movies and excited by the growing power of home computers had high expectations. For years, AI researchers had implied that a breakthrough was just around the corner. ("Within a generation the problem of creating 'artificial intelligence' will be substantially solved," Dr Minsky said in 1967.) Prototypes of medical-diagnosis programs, speech recognition software and expert systems appeared to be making progress. The 1985 conference of the American Association of Artificial Intelligence (AAAI) was, recalls Eric Horvitz, now a researcher at Microsoft, attended by thousands of people, including many interested members of the public and entrepreneurs looking for the next big thing.

It proved to be a false dawn. Thinking computers and household robots failed to materialise, and a backlash ensued. "There was undue optimism," says David Leake, a researcher at Indiana University who is also the editor of AI *Magazine*, which is published by the AAAI. "When people realised these were hard problems, there was retrenchment. It was good for the field, because people started looking for approaches that involved less hubris." By the late 1980s, the term AI was being eschewed by many researchers, who preferred instead to align themselves with specific sub-disciplines such as neural networks, agent technology, case-based reasoning, and so on. The expectations of the early 1980s, says Dr Horvitz, "created a sense that the term itself was overblown. It's a phrase that captures a long-term dream, but it implicitly promises a lot. For a variety of reasons, people pulled back from using it."

Ironically, in some ways, AI was a victim of its own success. Whenever an apparently mundane problem was solved, such as building a system that could land an aircraft unattended, or read handwritten postcodes to speed mail sorting, the problem was deemed not to have been AI in the first place. "If it works, it can't be AI," as Dr Leake characterises it. The effect of repeatedly moving the goal-posts in this way was that AI came to refer to blue-sky research that was still years away from commercialisation. Researchers joked that AI stood for "almost implemented". Meanwhile, the technologies that worked well enough to make it on to the market, such as speech recognition, language translation and decision-support software, were no longer regarded as AI. Yet all three once fell well within the umbrella of AI research.

Quiet respectability

But the tide may now be turning. "There was a time when companies were reluctant to say 'we're doing or using AI', but that's now changing," says Dr Leake. A number of start-ups are touting their use of AI technology. Predictive Networks of Cambridge, Massachusetts, focuses advertising using "artificial intelligence-based Digital Silhouettes" that analyse customer behaviour. The firm was founded by Devin Hosea, a former National Science Foundation fellow in artificial intelligence.

Another firm, HNC Software of San Diego, whose backers include the

Defence Advanced Research Project Agency in Washington, DC, reckons that its new approach to neural networks is the most powerful and promising approach to artificial intelligence ever discovered. HNC claims that its system could be used to spot camouflaged vehicles on a battlefield or extract a voice signal from a noisy background – tasks humans can do well, but computers cannot. HNC was acquired by Fair Isaac, another software firm, for $810m in 2002, and its technology is now used to analyse financial transactions and spot credit-card fraud.

Large companies are also using the term. Dr Leake points out that Bill Gates of Microsoft gave the keynote speech at the 2001 AAAI conference and demonstrated several Microsoft technologies that are close to being incorporated into the company's products. In February 2001, Microsoft trumpeted a "breakthrough application that enlists the power of artificial intelligence to help users manage mobile communications".

The product in question was Mobile Manager, which uses Dr Horvitz's research into Bayesian decision-making to decide which e-mail messages in an individual's in-box are important enough to forward to a pager. Dr Horvitz says he is happy to refer to his work as AI. His current work, which involves using spare computing capacity to anticipate and prepare for the user's most likely next action, is based on research published in *Artificial Intelligence*. "We just submitted a paper on how a theorem-proving program could exploit uncertainty to run more efficiently," he says. "That's core AI. I personally feel better about using the term. There are people, myself and others, who use the term proudly."

Sony also unabashedly uses the term AI when referring to its robot dog, AIBO. (The name is derived from the combination of "AI" and "bot", and means companion in Japanese.) The company boasts that "advanced artificial intelligence gives AIBO the ability to make its own decisions while maturing over time". It sounds like hype, though once you have seen an AIBO's uncannily life-like behaviour, the AI label seems appropriate. AIBO's intelligence, such as it is, relies on genetic algorithms, another trick that has been dug out from the AI toolkit.

In computer gaming, the term AI has always been used with a straight face. The gaming community got interested in AI in the late 1980s when personal computers started to get more powerful, says Steven Woodcock, a programmer who has worked in both the defence and games industries, and who maintains a website devoted to the study of AI in gaming: www.gameai.com. As graphics improve, he says, a game "needs other discriminators, like whether it plays smart". Game reviews routinely refer to the quality of the AI – well, what else would you call it? – and some games are renowned for the lifelike quality of their computer opponents.

Mr Woodcock says there is now quite a lot of traffic in both directions between AI programmers in the academic and gaming worlds. Military simulators, he notes, are increasingly based on games, and games programmers are good at finding quick-and-dirty ways to implement AI techniques that will make computer opponents more engagingly lifelike. Gaming has also helped to advertise and popularise AI in the form of such impressive games as *The Sims*, *Black & White* and *Creatures*.

Information overload

Another factor that may boost the prospects for AI is the demise of the dotcoms. Investors are now looking for firms using clever technology, rather than just a clever business model, to differentiate themselves. In particular, the problem of information overload, exacerbated by the growth of e-mail and the explosion in the number of web pages, means there are plenty of opportunities for new technologies to help filter and categorise information – classic AI problems. That may mean that artificial-intelligence start-ups – thin on the ground since the early 1980s – will start to emerge, provided they can harness the technology to do something useful. But if they can, there will be no shortage of buzzwords for the marketing department.

Not everyone is rushing to embrace this once-stigmatised term, however. IBM, for example, is working on self-healing, self-tuning systems that are more resilient to failure and require less human intervention than existing computers. Robert Morris, director of IBM's Almaden Research Centre in Silicon Valley, admits this initiative, called "autonomic computing", borrows ideas from AI research. But, he says, where AI is about getting computers to solve problems that would be solved in the frontal lobe of the brain, autonomic computing has more in common with the autonomic nervous system. To some extent, he suggests, the term AI has outgrown its usefulness. He notes that it was always a broad, fuzzy term, and encompassed some fields whose practitioners did not regard their work as AI. And while IBM continues to conduct research into artificial intelligence, Dr Morris does not link autonomic computing to such work. "This stuff is real," he says.

Similarly, Max Thiercy, head of development at Albert, a French firm that produces natural-language search software, also avoids the term AI. "I consider the term a bit obsolete," he says. "It can make our customers frightened." This seems odd, because the firm's search technology uses a classic AI technique, applying multiple algorithms to the same data, and then evaluates the results to see which approach was most effective. Even so, the firm prefers to use such terms as "natural language processing" and "machine learning".

Perhaps the biggest change in AI's fortunes is simply down to the change of date. The film A.I. was based on an idea by the late director, Stanley Kubrick, who also dealt with the topic in another film, *2001: A Space Odyssey*, which was released in 1969. *2001* featured an intelligent computer called HAL 9000 with a hypnotic speaking voice. As well as understanding and speaking English, HAL could play chess and even learned to lip-read. HAL thus encapsulated the optimism of the 1960s that intelligent computers would be widespread by 2001.

But 2001 has been and gone, and there is still no sign of a HAL-like computer. Individual systems can play chess or transcribe speech, but a general theory of machine intelligence remains elusive. It may be, however, that now that 2001 turned out to be just another year on the calendar, the comparison with HAL no longer seems quite so important, and AI can be judged by what it can do, rather than by how well it matches up to a 30-year-old science-fiction film. "People are beginning to realise that there are impressive things that these systems can do," says Dr Leake hopefully. "They're no longer looking for HAL."

The material on pages 336–40 first appeared in *The Economist* in March 2002.

Index